PROBLEMS

how to Solve them

and

Volume 3

The CENTRE for EDUCATION in MATHEMATICS and COMPUTING
Faculty of Mathematics
University of Waterloo
Waterloo, Ontario, Canada
1999

Published by
The CENTRE for EDUCATION in MATHEMATICS and COMPUTING
Faculty of Mathematics
University of Waterloo
Waterloo, Ontario, Canada

Telephone: (519) 885-1211, extension 3030
Fax: (519) 746-6592

Canadian Cataloguing in Publication Data

Main entry under title:

Problems and how to solve them : volume 3

ISBN 0-921418-14-0

1. Mathematics—Problems, exercises, etc. I. Anderson, Edwin.
II. University of Waterloo. Centre for Education in Mathematics and
Computing.

QA43.P76 1999   510'.76   C99-930369-4

Printed by Graphic Services, University of Waterloo

Authors

Edwin Anderson
Waterloo Mathematics Foundation
University of Waterloo

Enzo G. Carli
Centre for Education in Mathematics and Computing
University of Waterloo

Ronald G. Dunkley
Centre for Education in Mathematics and Computing
University of Waterloo

J. Stuart Telfer
Waterloo County Board of Education

Foreword

This book is about mathematical problem solving. It is the third in a series aimed at helping students, their teachers, and other interested persons gain a greater understanding of this extremely important area of mathematical study. While the first two books were designed for secondary school students, this one addresses the mathematics of grades 7 and 8.

The book includes discussion on topics contained in the curricula of all jurisdictions. As well as basic ideas, extensions are provided that will help both students and teachers better understand the basics on which the subject is built. There are numerous worked examples, exercises designed to ensure that fundamentals are understood, and problems ranging from very simple to demanding. Answers to the exercises, including solutions in some cases, are given, as well as complete solutions to all the problems.

The final chapter consists of miscellaneous problems which are useful for those who wish to test their problem solving skills without having the problems categorized by topic. Students, particularly those who are gifted in mathematics or who enjoy a challenge, should be encouraged to attempt exercises and problems on a regular basis. Since many of the questions are adapted from problems taken from the annual Gauss Mathematics Contests for Grades 7 and 8 students, the book can serve as a useful preparation guide for those contests.

In fact, much of the material could be used for students in the lower grades in high school to review and extend material taken earlier, as well as to prepare for the Pascal and Cayley mathematics contests. Parents are frequently asked to help their children solve mathematics problems. The authors feel that the book can serve as a valuable resource for this purpose. Finally, it is our hope that the book will be used by teachers groups taking special courses on teaching mathematics and during professional development days.

The authors are particularly indebted to Bonnie Findlay, University of Waterloo, for the cover design and for the many hours of typesetting she provided. Her skills in drawing diagrams, page set-up and printing procedures were extremely valuable to us. In particular her cheerful patience in handling a never ending stream of revisions is much appreciated.

The authors gratefully acknowledge the encouragement and support of the Faculty of Mathematics at the University of Waterloo in the production of this book.

E.A., E.G.C., R.G.D., J.S.T.
February, 1999.

Contents

Chapter 1   Introduction to Problem Solving

Problem solving is an ability people need in all areas of life. This book is written to help you solve mathematical problems. What, may you ask, is a mathematical problem? Well, try the following question.

Suppose you were asked to compete the following instructions.

1.   Select a number.
2.   Add 3 to the number.
3.   Multiply the result in step 2 by 4.
4.   Subtract 10 from the result in step 3.
5.   Divide your answer in step 4 by 2.
What is the result?

Would this be a problem for a student in grade 7 or 8? Probably not, since all you have to do is follow the instructions to get the result. But suppose that another student who followed these instructions obtained the result 21, and you are asked what was her original number. Now I think that you have a problem.

How could you solve this problem?

Since the answer is 21, you might make some guesses, check, and then refine your estimate until you get 21. For example, if you choose to start with 8, the result is 17. You are getting close to 21. Now start with a number larger than 8.

Another way to begin is to organize the numbers in a table in an attempt to establish a pattern. For example, if you begin with two the result is five, four gives an answer of nine, and six gives an answer of thirteen. We can show these results in a table.

Starting Number	Result
2	5
4	9
6	13

1

Note that for an increase of two in the starting number, the result increases by
4. Based on this observation, you could predict that starting with 8 gives a
result of 17 and starting with 10 would give a result of 21.

You might work backwards by using the inverse (or opposite) operations.

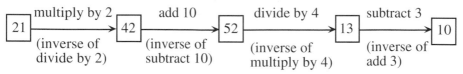

Let's check our answer.

In this problem we found three different ways to come up with an answer.
This tells us that a solution is usually possible if you are persistent and willing
to try different techniques or strategies.

If a problem does not have an automatic or obvious solution when we try to
solve it, then it is a mathematical problem. In this book we have both
exercises and problems. Exercises are routine type questions in which the
approach is apparent. The questions in the problem sections require the use of
many skills, the willingness to take a risk and a certain amount of creativity.
Sometimes you might find out that your approach leads to a dead end and no
solution is evident. There is nothing wrong with this. You simply have to try
a different approach. Sometimes your effort might lead to different solutions
or might even suggest other problems.

We start by looking at a few examples and making a list of strategies that
might help in solving problems. Remember that knowing a list of possible
steps is only a start. The thing that makes problem solving interesting is the
fact that we have to be able to create new strategies. There is always a need
for new ideas. As you solve problems you will come up with your own
techniques. One sure way of getting ideas for solving problems, when you
get stuck, is to talk to your teacher or to your friends.

Example 1

Place one or more of the arithmetic operations +, - , ×, and ÷ between each pair of successive numbers in the list 1, 2, 3, 4, 5, 6, 7, 8, and 9 so that the result is 100.

Solution

This is a mathematical problem because the answer requires some thought and creativity.

Step 1. Read the problem carefully and restate it in your own words to make sure you understand what is being asked.

Step 2. Collect all the information you can.
To solve this problem we need to know how to add, subtract, multiply, divide, and the order of operations.

Step 3. Explore different ideas. For example, break the problem into manageable parts.

Let's begin by adding some of the numbers: $1+2+3+4+5+6+7 = 28$. All we need is 72 more to make 100 and $8 \times 9 = 72$. Eureka! We have a solution.

Step 4. Solve the problem by putting our facts together.

$$1+2+3+4+5+6+7+8 \times 9 = 100$$

Can we come up with another solution?

Let's start by adding all the numbers from 1 to 9. The answer is 45.
Let's multiply some numbers: $4 \times 5 = 20$
$$7 \times 8 = 56.$$
Add: $20 + 56 = 76$.

We need 24 more to make 100. Can we make 24 out of 1, 2, 3, 6, and 9?
We know that $24 = 12 + 12$ and also that $12 = 6 \times 2$ or $3 + 9$.
We can also write 12 as $1 \times 2 \times 3 + 6$ and this includes the 1.

Let's try it. $1 \times 2 \times 3 + 4 \times 5 + 6 + 7 \times 8 = 88$. We need 12 more and we only have the number 9 to use. This is not working; we have come to a dead end.

Let's keep trying. We can see that $1 \times 2 \times 3 + 4 \times 5 + 6 + 7 \times 8 + 9 = 97$. If we "play around" with these numbers, we get $1 \times 2 \times 3 \times 4 + 5 + 6 + 7 \times 8 + 9 = 100$. We did it again!

Another solution is $1 - 2 + 3 \times 4 \times 5 + 6 \times 7 + 8 - 9 = 100$.

Now that we have seen two problems and worked out solutions, let's summarize some of the hints and strategies that we could try in solving mathematical problems.

- Read and understand the problem. Make sure that you are clear about what you are asked to do.
- Write down all the facts and the key words.
- Experiment with different numbers. In the second example, we kept trying different combinations of the numbers in order to get the answer 100.
- Make a table, a diagram, or a model. If you use a diagram, make a large one and mark all given facts and any other conclusions you can make on it.
- Look for a pattern. In example 1, we discovered a pattern in the table.
- Guess and check. You will be surprised how often this will lead to a solution.
- Don't panic. If what you have tried doesn't work, discuss the problem with your teacher or your friends. But don't ask for a solution. Ask for a hint and then try the problem again.
- Solve a simpler problem.
- Once you have a solution, write it down neatly, with all the steps explained thoroughly.
- Check your answer. Does it make sense?
- Keep practising. To become a good problem solver, you need to practise and solve different types of problems. The more problems you try, the better you will get.

Example 2

How many posts are required to support a straight fence that is 300 m long, if the posts are 5 m apart?

Solution

Let's look at a simpler problem, where we have 100 m of fence and the posts are 10 m apart.

Draw a diagram to help with the solution.

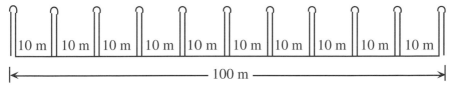

For 100 m of fencing we need 11 posts. (Note: this is $100 \div 10 + 1 = 11$ posts.) Since we have 300 m of fencing and the posts are 5 m apart, we need $300 \div 5 + 1 = 61$ posts.

Example 3

You are given the set of numbers 9, 18, 21, 27, 36, and 48. Does any subset of numbers from this set have a sum of 100?

Solution

Let's try some of these numbers.

$$48 + 21 + 9 + 36 = 114$$
$$48 + 36 + 27 = 111$$
$$48 + 27 + 21 + 9 = 105$$
$$9 + 18 + 21 + 48 = 96.$$

But wait a minute! Notice that each of the numbers in the list is a multiple of 3.

Let's examine the four results above. Are each of the answers multiples of 3?

$$114 \div 3 = 38, \ 111 \div 3 = 37, \ 105 \div 3 = 35, \text{ and } 96 \div 3 = 32.$$

Yes, they are all multiples of 3.

Let's try another sum: $9 + 18 + 21 = 48$ and $48 \div 3 = 16$, so 48 is a multiple of 3.

Also, $9 + 27 + 36 + 4 = 120$ and 120 is a multiple of 3.

It is true, in fact, that if the numbers we add are multiples of 3, then their sum is a multiple of 3.

But 100 is not a multiple of 3, since $100 \div 3 = 33.33333\ldots$.

Therefore no subset of the given numbers has a sum of 100.

Is it possible for a subset of the numbers 14, 21, 42, 49, and 63 to add to 100? Since all the given numbers are multiples of 7, and 100 is not a multiple of 7, then it is not possible for any subset of these number to add to 100.

As a check, let's try a few: $14 + 21 + 63 = 98$ and $98 \div 7 = 14$.

You try some more.

Is it possible for a given subset of the numbers 14, 21, 42, 49, and 63 to add to 105?

Check if 105 is a multiple of 7. Since $105 \div 7 = 15$, it might be possible.

For example $42 + 63 = 105$. You check some other possibilites.

Example 4

Each visible face of a block in the
pile shown measures 2 cm by 8 cm.
Determine the length of the path
marked with the solid line segments.

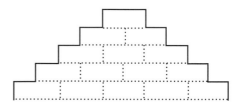

Solution

The path consists of upward, downward and horizontal distances.

The sum of the upward distances is $5 \times 2 = 10 \, \text{cm}$.

The sum of the downward distances is also 10 cm.

All the horizontal distances add up to the length of the bottom line.

Since the length of the bottom line is $5 \times 8 = 40 \, \text{cm}$, the length of the path is $10 + 10 + 40 = 60 \, \text{cm}$.

Example 5

Each of the numbers 1, 2, 3, 4, 5, and 6 is to be placed in one of the boxes in the diagram. Each number below two adjacent numbers must be equal to their *positive* difference. This means, for example, that adjacent boxes can have 3 and 2 or 2 and 3 to give a difference of 1.

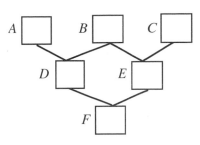

What is the largest number that can be placed in *F*, the bottom box?

Solution

We are looking for the largest number that can be placed in box *F*. The largest number from the given list is 6. If 6 is placed in box *F*, none of the other five numbers have a difference of 6, therefore 6 is not possible in position *F*.

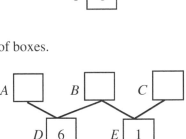

For the same reason, 6 cannot be placed in either box *D* or box *E* .

Therefore 6 can only be placed in the top row of boxes.

If we place 5 in box *F*, then 6 and 1 must be placed in the second row. But we have already noted that 6 must be placed in the first row, so 5 cannot be placed in box *F*.

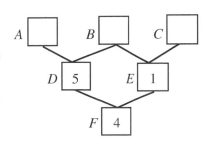

Let's consider 4 in box F.

We can place 5 in box *D*, and 1 in box *E*, or we can place 6 in box *D* and 2 in box *E*. But 6 cannot be placed in the second row of boxes. (Why?)

Therefore place 5 in box *D*, and 1 in box *E*. To get a difference of 5, we need to place a 6 and a 1 in the first row. But we have already used 1 in box *E*.

Then 4 is not possible in box *F*.

Consider 3 in box F.

The possibilities for boxes D or E are

(i)   6 in D and 3 in E; but this is not possible because 6 must be placed in row 1.

(ii) 5 in D and 2 in E; therefore 6 must be placed in box A or B. If we place 6 in box A and 1 in box B, then we must place 3 in box C. But this is not possible since we used 3 in box F. Therefore the top row must have 1 in box A, 6 in box B, and 4 in box C.

(iii) 4 in D and 1 in E . Try this yourself to see that there is a second way of doing this.

Therefore 3 is the largest number that can be placed in box F .

From the examples and discussion in this section you have seen several approaches to problem solving. A summary of the main steps in problem solving will help you bring the skills together.

1.   *Identify the problem.*
     Read the problem carefully several times. What do you have to find?

2.   *Determine the information required.*
     What information or materials are needed before you can solve the problem?

3.   *Obtain the information.*

4.   *Solve the problem.*

5.   *Check your answer and write a final statement.*

Problems

1. A piece of string is 72 cm long. Where must it be cut for one piece to be three times as long as the other piece?

2. How many posts are required to support a straight fence that is 236 m long, where the posts are 4 m apart?

3. Do any combination of the numbers 10, 20, 30, 40, and 50 add to 75? If no, why not?

4. If a tap drips at the rate of 3 drops every 5 seconds, how many drops will fall in one minute?

5. Two elevators leave the sixth floor of a building at the same time. The faster elevator takes 10 seconds between floors and the slower one takes 20 seconds between floors. The first elevator to reach a floor must stop for 30 seconds to pick up passengers. The other elevator does not stop if passengers have been picked up. Which elevator reaches the ground floor first if there are passengers to be picked up on every floor?

6. In the dotted rectangle *ABCD*, $AB = 20$ cm and $BC = 18$ cm. If all the angles in the diagram are right angles, what is the length of the solid line from *A* to *D*?

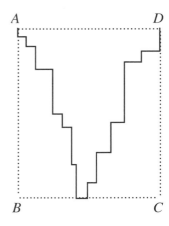

7. Determine the distance around the outside of the given figure.

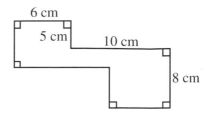

8. Place the integers 1 to 6 in the six circles so that adjacent pairs, like 1 and 2, or 4 and 5, go in circles which are not joined by a line.

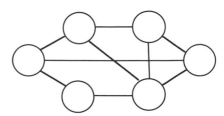

9. Place the integers 1 to 8 around a square, one at each vertex and one on each side, such that the sum of the numbers along each side is the same?

10. Rank the heights of the given students, given that
    (i)   Jane is taller than Sandy
    (ii)   Sandy is not as tall as Ed
    (iii)   Alicia is taller than Sandy but shorter than Ed
    (iv)   Ron is not as tall as Alicia, but taller than Sandy
    (v)   Gurinder is not as tall as Jane, but taller than Ed.

11. In the diagram, all triangles have sides of length 1. What is the length of the longest path from A to B, without going along any side of a triangle more than once?

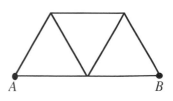

12. In the diagram, each of the seven circles is coloured so that no two circles connected by a line segment have the same colour. What is the minimum number of different colours required?

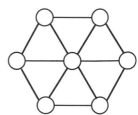

13. Draw two straight lines in the circle such
     that the sum of the numbers in each of the
     three areas is equal.

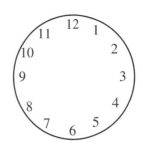

14. Each of the integers 1 to 10 is to be
     placed in one of the boxes in the
     diagram. Each integer below two
     adjacent integers equals their ***positive***
     difference.
     Given 4 and 6 in the boxes shown,
     finish placing the remaining integers in
     the other boxes.

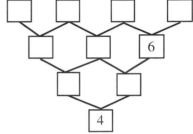

Chapter 2   Numeracy

2.1   Prime Numbers

Throughout the ages there has been a fascination with numbers. Everyone learns to count at an early age and we are familiar with the counting numbers that we designate as *whole numbers.*   A subset of the whole numbers, referred to as the prime numbers was of great interest to mathematicians more than 2000 years ago. *A prime number is a positive integer greater than 1 that has only two divisors, 1 and itself.* For example 2 is a prime number .   Other prime numbers are 3, 5, 7, and 11.   The number 4 is a *composite number,* because it is divisible by three positive integers, 1, 2 and 4. By agreement, the number 1 is neither a prime nor a composite number.
Is 17 a prime or a composite number? How about 18?

The Greek mathematician Eratosthenes devised a technique that allows us to find all the prime numbers smaller than a given number.   This method is called the "sieve of Eratosthenes", and gives the primes as an ordered set.

Suppose you want all the prime numbers less than 50. First, list all the numbers between 2 and 50 inclusive.

Begin by circling the 2 and cross out all multiples of 2. (None of them can be prime. Why?)

Now circle the first surviving number, 3, and cross out all multiples of 3. (Some have been removed already. Why?)

Now circle the next surviving number, 5, and cross out all multiples of 5. Then circle 7 and cross out all multiples of 7. If you continue this process until you reach 50, every *circled* number will be a prime.

In fact, we stopped circling numbers at 23.   Any number larger than one-half of 50, and which remains uncircled cannot have a multiple less than 50, so all uncircled numbers from this point on must be prime.

Suppose we want to determine whether a number, for example 73, is a prime without listing all the numbers less than or equal to the given number? From above we know that there is no need to try any number larger than half of 73, since any number greater than 37 cannot have a multiple less than 73. In fact, we can do much better than that, using the following logic.

If a number is composite, then it can be expressed as the product of two factors, other than one and itself. These factors might themselves be composite and each will be smaller than the original number. Unless the number is a perfect square, one of these factors is smaller than the square root of the number, the other larger. (Why must this be?) For example,

$54 = 6 \times 9$ and $\sqrt{54} \doteq 7.348$ ($\doteq$ means approximately equal to).

In testing 73, we note that 73 could have factors a and b, where a is less than $\sqrt{73} \doteq 8.544$, and b is greater. Hence a must be 2, 3, 4, 5, 6, 7, or 8. We need not test all of these, because if 4, 6, or 8 divides 73, then surely 2 does, so it is sufficient to test 2, 3, 5, and 7, the prime numbers less than $\sqrt{73}$. Isn't this remarkable! We need test only four numbers, and if none of them divides 73, then it is a prime number. You can determine immediately without even trying that 2 and 5 are not factors of 73. (Why?) So you only have to try 3 and 7. Since neither 3 nor 7 divides 73, then 73 is a prime number.

Note: When we say that 7 divides 28, we mean that 7 divides 28 exactly .

Let's try this process with another number just to make sure that we have it.

Example 1
Is 133 a prime number?

Solution

Step 1. Find the square root of 133.	$\sqrt{133} \doteq 11.53$
Step 2. List the prime numbers less than or equal to 11, the whole number part of $\sqrt{133}$.	The prime numbers less than $\sqrt{133}$ are 2, 3, 5, 7, and 11.
Step 3. Check if any of the primes less than $\sqrt{133}$ divide exactly into 133.	Check if 133 is divisible by 2, 3, 5, 7, or 11. 7 divides 133.

Since 7 divides 133, then 133 is not a prime.

It is true that with relatively small numbers we can determine whether or not they are prime by checking mentally. The system described is useful with larger numbers.

Example 2
Is 487 a prime number?

Solution
$\sqrt{487} \doteq 22.07$
The prime numbers less than or equal to 22 are 2, 3, 5, 7, 11, 13, 17, and 19. Since 487 is not divisible by any of these, then 487 is a prime number.

Exercises
1.   When was Eratosthenes born? Eratosthenes made another beneficial contribution to mathematics. What was it?

2.   Use the sieve of Eratosthenes to determine all the primes less than 100. List these prime numbers.

3.   Use the "square root method" to determine if the following numbers are prime.
      (a) 593   (b) 921   (c) 2789

4.   Is the year of your birth a prime number? Is your address a prime number?

5.   If you are a computer buff you might be interested in writing a computer programme that determines whether or not a number is prime.

2.2   *Prime Factors*

Prime numbers are fundamental to solving many problems in mathematics, so knowing how to factor a number into its prime factors can be very useful. The operation of factoring is the reverse of multiplication. For example, $4 \times 3 = 12$ and this is the only answer possible. The reverse process called factoring asks for numbers that can be multiplied together to give 12. This process gives more than one answer. We have listed several different ways of obtaining 12. These are 2×6, 12×1, 4×3, $(-3)(-4)$, $2\times2\times3$, 24×0.5, and $\frac{1}{3}\times36$.

Note that some of these combinations involve fractions, others negative numbers. In determining the factors of an integer, we agree that we will not consider order (3×4 is the same as 4×3) and we will use only positive integers. That is, we will not use $(-3)(-4)$.

With these agreements we have reduced our possible combinations to

$$12 = 1\times12$$
$$= 2\times6$$
$$= 3\times4$$
$$= 2\times2\times3.$$

Note that $12 = 2\times2\times3$ contains only prime numbers, so $2\times2\times3$ is the prime factorization of 12.

Example 1
Write the number 42 as the product of prime factors.

Solution
One of the easiest ways of finding the prime factors of a number is to try dividing it by each of the prime numbers in order, that is, by 2, 3, 5, and so on.

Note that 42 is divisible by 2 since $42\div2 = 21$ and 21 is divisible by 3 since $21\div3 = 7$.

The answer is $42 = 2\times3\times7$.

We can write these steps using a tree diagram as shown.

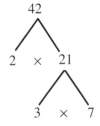

Here are some interesting facts that will assist in determining factors.
A number is divisible by 2 if the last digit is even (0, 2, 4, 6, 8).
A number is divisible by 3 if the sum of its digits is divisible by 3.
A number is divisible by 4 if the number represented by the last two digits is divisible by 4.
A number is divisible by 5 if the last digit is 0 or 5.
A number is divisible by 9 if the sum of its digits is divisible by 9.

Example 2

Write 252 as the product of prime factors.

Solution

In using the factor tree, begin with 2×126 since
252 is an even number and 2 is a prime number.
The same reasoning is used for factoring 126 into
2×63. The other factors you can read from the
diagram.

Therefore $252 = 2 \times 2 \times 3 \times 3 \times 7$.

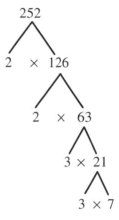

Of course if you had a number like 240, you could
begin with 10×24 since the last digit of 240 is 0,
and any number that ends in a zero has 10 as one of
its factors.

Use this method to find the prime factors of 240.

Exercises

6. Write each of the following numbers as the product of prime factors.

   (a) 63   (b) 70   (c) 123   (d) 1575

7. What are the two greatest prime factors of 5757?

Example 3

Make an expression equal to 100 using each of the numbers 1, 2, 3, 4, 5, and 6
exactly once. You may use only parentheses and any of the four basic
arithmetic operations.

Solution

In solving this problem, the first thought that comes to mind is that the
number 100 can be broken down in a number of ways. For example,
$100 = 10 \times 10$. One approach to this problem is to make two tens using the
numbers 1 through 6.

One way of making ten is $6 + 4$. Now we need to use the remaining numbers
1, 2, 3, and 5 to make 10. Since the sum of $5 + 3$ is two less than 10, then we
need to make a 2 out of the remaining numbers 1 and 2. The easiest way to do
this is to multiply 1×2.

We can summarize our solution using a tree diagram.

Therefore $100 = (6+4) \times (5+3+2 \times 1)$.

There are many other solutions to this problem.

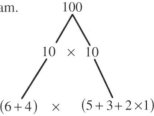

Example 4
Make an expression equal to 100 by using each of the numbers 1, 2, 3, 4, 5, and 6 exactly once. Here you may use exponents and square roots as well as parentheses and any of the four basic arithmetic operations .

Solution
100 can be written as $30 + 70$.

But $30 = 5 \times 3 \times \sqrt{4}$ and $70 = (6^2 - 1) \times 2$, since $6^2 = 6 \times 6$ and $36 - 1 = 35$.

Hence $100 = (5 \times 3 \times \sqrt{4}) + (6^2 - 1) \times 2$.

Exercises
8. Determine two other solutions to example 3.

9. Determine two other solutions to example 3, only this time begin with the product 20×5.

10. Determine two other solutions to example 3, but this time begin with a product other than 10×10 or 20×5.

11. Determine two other solutions to example 4 using parentheses, any of the four basic operations, exponents, and square roots.

12. Determine two ways of making an expression equal to 1000, using each of the numbers 2, 3, 4, 6, and 9 exactly once, any of the four basic operations, parentheses, exponents, and square roots.

Can you find additional solutions to these exercises?

2.3   *Divisors and factors*

The number 10 is divisible by 2, since $10 \div 2 = 5$, and 5 is an integer.
Since $10 = 2 \times 5$, all possible divisors of 10 are 1, 2, 5, and 10. There are 4 divisors of 10.
Note that the factors of 10 are 2 and 5 (since $10 = 2 \times 5$), and 10 and 1 (since $10 \times 1 = 10$).

Example 1
What are the divisors of 18?

Solution
First write the prime factors of 18, that is, $18 = 2 \times 3 \times 3$.
The six divisors of 18 are 1, 2, 3, 6, 9, and 18.

Note that the factors of 18 are :
$$18 = 1 \times 18$$
$$= 2 \times 9$$
$$= 3 \times 6$$
$$= 2 \times 3 \times 3.$$

Example 2
The three digits 4, 5 and 9 form three-digit numbers such as 549. How many of these three-digit numbers are divisible by 6?

Solution
For a number to be divisible by 6, it must be divisible by 2 since $6 = 2 \times 3$ and therefore it must be even. It must also be divisible by 3 and a number is divisible by 3 if the sum of its digits is divisible by 3. The sum of the digits 4, 5 and 9 is 18, which is divisible by 3. The only even numbers using these digits are 594 and 954, both of which are divisible by 6.

Example 3
Find the smallest natural number by which 132 can be multiplied so that the number created is divisible by 24.

Solution
Since $24 = 2 \times 2 \times 2 \times 3$ and $132 = 2 \times 2 \times 3 \times 11$, we need to multiply 132 by 2 to create a number divisible by 24.

Check: $132 \times 2 = 264$ and $264 \div 24 = 11$.

Example 4
What is the smallest natural number by which 54 can be multiplied so that the result is a perfect square?

Comment
Before we can write a solution to Example 4, we should consider the prime factors of a perfect square.

For a perfect square, each prime factor must occur an even number of times. For example, 25 is a perfect square and $25 = 5 \times 5$; the prime factor 5 occurs twice. Another perfect square is 36 and $36 = 2 \times 2 \times 3 \times 3$. Note that the prime factors 2 and 3 each appear twice, or an even number of times.

Find the prime factors of each of the following perfect squares.
(a)   49   (b)   144   (c)   225   (d)   2500

How many times does each prime factor appear in each perfect square? What conclusion can you make about the number of times each prime factor appears in each perfect square?

Solution to example 4
First write the prime factors of 54. Since $54 = 2 \times 3 \times 3 \times 3$, both the 2 and the 3 occur an odd number of times. If we multiply 54 by both 2 and 3, then the prime factors 2 and 3 will occur an even number of time.

$$54 \times \boxed{2 \times 3} = 2 \times 3 \times 3 \times 3 \times \boxed{2 \times 3}$$

Therefore the smallest natural number by which 54 can be multiplied in order to get a perfect square is 2×3 or 6.

Check: $54 \times 6 = 324$ and $\sqrt{324} = 18$.

Exercises
13.   What is the smallest natural number 24 must be multiplied by so that the result is a perfect square?

14.   Find the smallest natural number that 94 must be multiplied by so that the number created is divisible by 12.

15. (a) If a number is divisible by 3 is it divisible by 9?
    (b) If a number is divisible by 9 is it divisible by 3?

16. Find the divisors and factors of each of the following numbers.
    (a) 24   (b) 27   (c) 52

2.3   *Greatest Common Factor (GCF) and the Lowest Common Multiple (LCM)*

In this section we discuss two important and useful mathematical concepts, illustrating them by examples. First consider this problem.

The Math Club donates 48 hot dogs and 84 cans of pop for a picnic. The club stipulates that each child is to receive the same amount of pop and hot dogs and all the pop and hot dogs have to be used. What is the largest number of students that can be invited to the picnic?

To answer the question, we need to find a number that divides both 48 and 84. Well, 2 divides them, so we could invite two students, and they could both get sick. We can see that 3 divides them, and so does 4, but if we want to invite the largest number possible, then we'll have to organize our thinking. Let's find the prime factors of 48 and 84.

$$48 = 2 \times 2 \times 2 \times 2 \times 3$$
$$84 = 2 \times 2 \times 3 \times 7$$

Each number contains, along with other things, two 2's and one 3, so each is divisible by $2 \times 2 \times 3$ or 12 and now we see that we could invite 12 students and give each of them 4 hot dogs and 7 cans of pop. (So somebody is still likely to get sick.)

Now we can see what the **Greatest Common Factor** (called the GCF for short) is.

The **Greatest Common Factor** (GCF) of a set of integers is the largest positive integer that divides exactly into each of the integers of the set.

Exercises

17. Find the GCF of each of the following groups of numbers.

(a) 28, 42　　　　　(b) 30, 45　　　　　(c) 84, 126, 210

18. What is the least positive integer that is exactly divisible by both 24 and 30?

Along with the GCF, we frequently use the **Lowest Common Multiple** (called LCM for short).

The **Lowest Common Multiple** of a set of two or more integers is the smallest positive integer that each of the integers in the set divides exactly.

Example 1

Find the LCM of 30 and 45.

Solution

Once again, let's use the prime factors of these numbers, that is,

$30 = 2 \times 3 \times 5$
and $45 = 3 \times 3 \times 5$.
For the LCM, we choose the largest number of times each factor occurs. In this case 2 occurs once, 3 occurs twice and 5 occurs once.
The LCM is $2 \times 3 \times 3 \times 5 = 90$.

The most common application of the LCM is in finding the lowest common denominator when we are adding and subtracting fractions.

Example 2

Evaluate $\dfrac{3}{4} + \dfrac{1}{8} - \dfrac{5}{12}$.

Solution

Since $4 = 2 \times 2$, $8 = 2 \times 2 \times 2$ and $12 = 2 \times 2 \times 3$, the LCM of 4, 8 and 12 is $2 \times 2 \times 2 \times 3 = 24$.

Then $\dfrac{3}{4} + \dfrac{1}{8} - \dfrac{5}{12} = \dfrac{18}{24} + \dfrac{3}{24} - \dfrac{10}{24}$

$$= \dfrac{11}{24}.$$

Example 3

Peter and Ron jump into the deep end of a pool at the same time and begin to swim lengths. Peter swims one length every 20 seconds and Ron swims one length every 12 seconds. How many lengths will Ron have swum at the moment when he and Peter reach the same end of the pool?

Solution

In order to solve this problem we need to find the smallest positive integer that 20 and 12 will divide exactly; that is, the LCM of these two numbers. First find the prime factors of 20 and 12.

Since $20 = 2 \times 2 \times 5$ and $12 = 2 \times 2 \times 3$, the LCM is $2 \times 2 \times 5 \times 3 = 60$.

Since $60 \div 12 = 5$, Ron and Peter meet at the same end of the pool when Ron has swum 5 lengths and Peter has swum 3 lengths. (Why?)

Exercises

19. Find the LCM of each of the following groups of numbers.

    (a) 28, 42   (b) 12, 30   (c) 84, 126, 210

20. What is the value of $\frac{1}{5} \times \frac{1}{2} + \frac{3}{4} - \frac{3}{8}$?

21. A student is given a supply of rectangles each measuring 24 mm by 39 mm. She wants to lay them out to form a square. Each rectangle is laid down with the long side horizontal.

    (a) What are the dimensions of the smallest possible square?
    (b) How many rectangles will be required to form the smallest square?

Example 4

A baseball team won 50 of its first 75 games. The team still has 45 games to play. For the team to win 60% of its games over the entire season,

(a) how many of the remaining games must it win?

(b) what percentage of its remaining games must the team win?

Solution

(a) The team plays a total of $75 + 45 = 120$ games for the season.

   60% of 120 is $0.60 \times 120 = 72$.

   The team has won 50 games so far.

   It must win $72 - 50 = 22$ of its remaining games.

(b)   To find this percentage divide 22 by 45 and multiply by 100.

$\frac{22}{45} \times 100 = 48.9$ (correct to one decimal place).

The team must win 48.9% of it remaining games.

The following problems require intuitive thought as well as various facts that you have learned.

Problems

1.   The six positive divisors of 12 are 1, 2, 3, 4, 6, and 12.
     (a)   What is the next largest number that has exactly six positive divisors?
     (b)   What are the divisors of the number you found in part (a)?
     (c)   What are the factors of the number you found in part (a)?

2.   The number 315 can be written as the product of two integers each greater than 1. In how many ways can this be done?

3.   The number 7315 can be written as the product of a pair of two-digit numbers. What are these numbers?

4.   Three brass numbers, 3, 5 and 7, are purchased from a hardware store.
     (a)   How many different two-digit house numbers can be formed using these numbers ?
     (b)   How many of these numbers are divisible by 6? Why?

5.   What is the value of
     $1 + 2 - 3 - 4 + 5 + 6 - 7 - 8 + \cdots + 45 + 46 - 47 - 48 + 49$?

6.   At the beginning of the week, a stock was worth $100. On Monday, its price went up 10%. On Tuesday, its price dropped 10% and on Wednesday its price went back up 10%. What is the percent increase in the price of the stock over the three day period?

7.   Given that the six-digit number $5913d\,8$ is divisible by 12, what are the possible values of d?

8. Three cards numbered 1, 2 and 3 are placed, one to a box, in the equation shown.

    $$n = \boxed{} + \boxed{} - \boxed{}$$

    How many different answers can be obtained?
    How many different answers are possible if you select from the numbers 1, 2, 3, and 4?

9. I gave $\frac{1}{3}$ of a pizza to my brother and another $\frac{1}{3}$ of the pizza to my sister.

    I gave $\frac{3}{4}$ of the remaining piece to my friend. What fraction of the pizza is left for me?

10. What is the value of the expression
    $(4+8+12+16+\cdots+256)-(1+5+9+13+\cdots+253)$?

11. The capacity of the gas tank in Kathy's car is 48 litres. When the tank is one-third full, how many litres must be added to make the tank three-quarters full?

12. What is the units (or ones) digit of the product
    $1 \times 3 \times 5 \times 9 \times 11 \times 13 \times 15 \times 17 \times 19 \times 21 \times 23$?

13. Mr. Metcalf loses his place when the book he is reading accidentally closes. He remembers that the sum of the two page numbers showing was 233. What is the product of the page numbers showing?

14. The three regions on a dart board are assigned scores of 3, 5 and 7, as shown. Stu hits the board with three darts. How many different scores could he have had?

    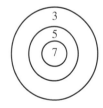

15. Don has four teenage children, each with a different age. The product of their ages is 67 184. What are their ages?

16. Given that $6+7+8+9+\cdots+482+483=116\,871$, what is the sum of $8+9+10+11+\cdots+484+485$?

17.   In each horizontal row of this grid, the three numbers have a different sum. By exchanging only two numbers, all three rows will have the same sum. What are these two numbers?

16	8	22
23	4	12
11	27	15

18.   The numbers 2, 3, 4, 5, and 6 are placed, one to a square, in the diagram shown. The sum of the numbers in the row and in the column must each be 13. What is number in the middle square?

19.   Chris was asked to add 8 to a given number and then divide by 2. Instead, Chris added 2 to the given number, then divided by 8 and obtained the answer 4. If Chris had followed the instructions correctly, what would have been the final answer?

20.   Each of the digits 3, 5, 6, 7, and 8 is placed, one to a box, in the diagram. If the two-digit number is subtracted from the three-digit number, what is the smallest difference?

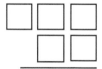

21.   The sum of five consecutive positive integers is 105. What is the largest of these integers?

22.   A bank employee is filling an empty cash machine with bundles of $5.00, $10.00 and $20.00 bills. Each bundle has 100 bills in it and the machine holds 10 bundles of each type. What amount of money is required to fill the machine?

23.   The figure shown is folded to form a cube. Three faces meet at each corner. If the numbers on the three faces at a corner are multiplied, what is the largest possible product?

	5	
	6	
7	3	2
	8	

24.   A bowl contains 40 g of white rice and 60 g of brown rice. If 100 g of white rice is added to the mixture, what percentage of the new mixture is white rice?

Chapter 3   Geometry

3.1   Lines and angles

Geometry is the study of the relationship between lines, angles and shapes. Some of these shapes appear in the world around us. Stars in the sky suggest points, the full moon a circle, the horizon between the sea and the sky a line. Some floor tiles are in the shape of rectangles or squares. Triangles are used for yield signs and eight-sided figures, called octagons, are used for stop signs.

The word geometry comes from the Greek word meaning "earth measure". More than 4000 years ago the Egyptians and Babylonians were using geometry in surveying and in architecture. These ancient mathematicians discovered geometric facts from experimentation and inductive reasoning. Geometry remains a rich source of problems that lend themselves to exploration and creative inquiry.

You are familiar with the distinctions between a straight line, a ray and a line segment, each of which is illustrated below. Because it is usually quite clear which of these is intended in any problem, we will normally use the term *line* to refer to any of them.

| line | ray | line segment |

With two lines we can form an angle.

An *acute angle* is an angle whose measure is less then 90°. We name an angle using letters. Here, for example, is an acute angle named ∠ABC. B is the vertex of the angle.

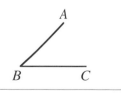

The line *AB* can pivot around the point *B* to create other angles as shown in the diagram.

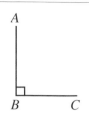

∠ABC is a right angle and has a measure of 90°. We also say AB is perpendicular to BC.

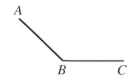

∠ABC is an obtuse angle and has a measure greater than 90° and less than 180°.

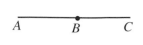

∠ABC is a straight angle and has a measure of 180°.

If two lines intersect, the opposite angles formed are equal.

$$\angle AEC = \angle DEB$$
$$\angle AED = \angle CEB$$

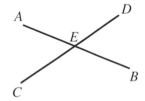

If two lines do not intersect, then they are parallel.

Here we write $AB \parallel CD$, meaning AB is paralled to CD, and we indicate this on a diagram by the arrows shown on the two lines.

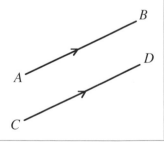

A *transversal* is a line that intersects two or more lines.

When a transversal, PT, intersects two parallel lines, AB and CD, as in the diagram,

(i)   alternate angles are equal, for example,
$$\angle BRS = \angle CSR$$

(ii)   corresponding angles are equal, for example, $\angle PRB = \angle RSD$.

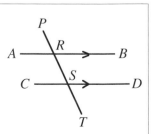

Example 1

What are the possible numbers of points of intersection for three lines in a plane?

Comment

Examples of planes in geometry are the surface of a pond on a calm day, a flat sheet of paper and the floor of a room.

Solution

Three lines in a plane may intersect in 0, 1, 2, or 3 points.

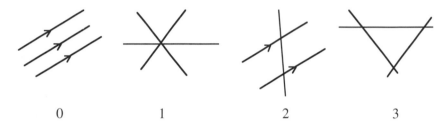

    0   1   2   3

Exercises

1.   What are the possible numbers of points of intersection for four lines in a plane?

2.   Use the diagram at the right to name two pairs of alternate angles and four pairs of corresponding angles.

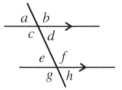

Example 2

Determine the value of b and then calculate the measure of $\angle DAC$.

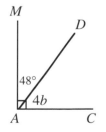

Solution

$\angle MAC = 90°$

$4b + 48 = 90$

$\quad 4b = 52$

$\quad\ b = 13.$

Therefore $\angle DAC = 4 \times 13°$

$\qquad\qquad\quad = 52°.$

Comment   *In this book , all angle measures represented by variables will be in degrees.*

Example 3
Determine the values of x and y.
Calculate the measure of each angle.

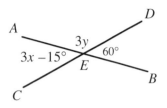

Solution
Since the lines intersect, the opposite angles are equal. That is,

$$\angle AEC = \angle DEB$$
$$3x - 15 = 60$$
$$3x = 75$$
$$x = 25.$$

$$3x - 15 = 3(25) - 15$$
$$= 60.$$

Since AEB is a straight line, $\angle AEB = 180°$.

$$3y + 60 = \angle AEB$$
$$3y + 60 = 180$$
$$3y = 120$$
$$y = 40.$$

Therefore $\angle AEC = \angle DEB = 60°$ and $\angle CEB = \angle AED = 120°$.

Note: The sum of the angles around E is 360°.

Example 4
Determine the values of x and y.

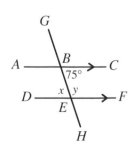

Solution
Since $\angle CBE$ and $\angle DEB$ are alternate angles between parallel lines,

$$\angle CBE = \angle DEB$$
$$x = 75.$$

Since $\angle DEF$ is a straight angle,

$$x + y = 180$$
$$75 + y = 180$$
$$y = 105.$$

Exercises

3.   Determine the value of each variable.

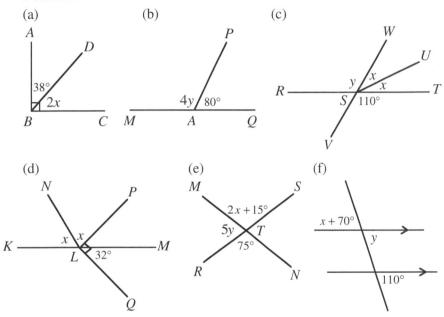

3.2   *Triangles*

Many of the basic shapes and solids we see are made up of polygons. A
polygon is a simple closed figure made up of line segments. In regular
polygons, all sides and all interior angles are equal.

A triangle is the simplest polygon. It has three sides and three interior angles.
In a triangle the sum of the measures of the three interior angles is 180°.

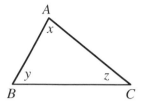

$$x + y + z = 180°$$

The diagrams below show why this relationship is true. Draw line PQ through A and parallel to BC. Since $\angle PAB$ and $\angle ABC$ are alternate angles, then $\angle PAB = y$. Similarly, $\angle QAC = z$.

Since $\angle PAB + \angle BAC + \angle CAQ = 180°$, then $x + y + z = 180°$.

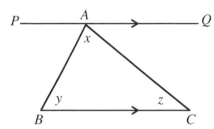 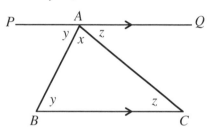

Triangles can be classified according to lengths of their sides.

Scalene	Isosceles	Equilateral
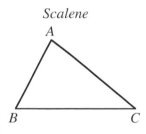	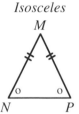	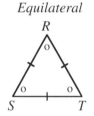
A *scalene* triangle has all three sides of different lengths.	An *isosceles* triangle has two sides equal. Here $MN = MP$. In an isosceles triangle the angles opposite the equal sides are equal. In this case $\angle MNP = \angle MPN$.	An *equilateral* triangle has three sides equal. This means that $RS = ST = RT$. The three angles in an equilateral triangle are equal. The measure of each angle is $60°$.

Example 1
Determine the value of b in the triangle.

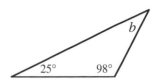

Solution
Since the sum of the angles in a triangle is $180°$, then

$$b + 25 + 98 = 180$$
$$b + 123 = 180$$
$$b = 57.$$

Example 2
Determine the values of x and y in the triangle.

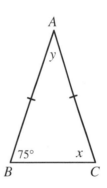

Solution
Since $AB = AC$, then $\triangle ABC$ is isosceles and so $x = 75$ (the two angles opposite the equal sides are equal).
In a triangle, the three angles add to $180°$.

$$y + 75 + 75 = 180$$
$$y + 150 = 180$$
$$y = 30.$$

Exercises
4. Calculate the value of each variable.

(a)

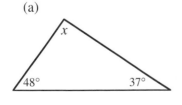

(b)

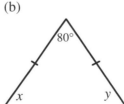

(c)

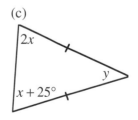

5.   In the diagram at the right, what is the sum of
     the acute angles at *A*, *B*, *C*, *D*, *E*, and *F*?

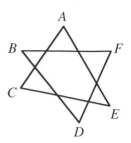

6.   In the diagram, the measure of angle *PRQ* is
     120° and the measure of angle *PST* is 110°.
     What is the measure of angle *RPS*?

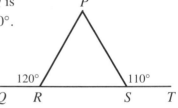

We can find the sum of the measures of the interior angles of a polygon by
using the fact that the sum of the interior angles of a triangle is 180°.

Consider a quadrilateral, a four sided polygon.

Take any point *P* inside the quadrilateral. Join *P* to
the four vertices, *A*, *B*, *C*, and *D*. Four triangles are
formed. Since the sum of the angles in a triangle is
180°, the four triangles inside the quadrilateral have
an angle sum of $4 \times 180° = 720°$.

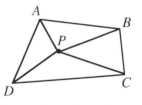

But the angles around point *P* are not part of the interior angles at the points
A, *B*, *C*, and *D*. Furthermore, the sum of the angles around *P* equals 360°.
The sum of the interior angles in a quadrilateral is $720° - 360° = 360°$.

Exercises

7.   Determine the sum of the measures of the interior angles of each of the
     following polygons.
     (a) Pentagon (five sides)   (b) Hexagon (six sides)
     (c) Septagon (seven sides)   (d) Octagon (eight sides)
     (e) Dodecagon (twelve sides).

8.   What is the sum of the interior angles of a 50-sided polygon? An
     n-sided polygon?

You should know that the sum of the lengths of two sides of a triangle is always greater than the length of the third side. This is called the *"Triangle Inequality"*.

In $\triangle ABC$, $AB + BC > AC$

$BC + AC > AB$

$AB + BC > AC$.

Try the following experiment to help you understand this idea.

Arrange three toothpicks to form a triangle. Continue the process with 4, 5, 6, up to 12 toothpicks. For each number of toothpicks decide whether any triangles are possible. If so, sketch the triangle and then determine whether other triangles can be made with this same number of toothpicks. The first few are done for you.

Number of toothpicks	Length of the sides of the triangle.	Sketch of the triangle.
3	1, 1, 1,	
4	not possible (why?)	
5	1, 2, 2	
6	2, 2, 2,	Draw the triangle.
7	1, 3, 3 There is another triangle possible. Find it.	Draw the two triangles.
9	2, 3, 4 Determine all the other possible triangles and then sketch them.	

Complete this table for 8, 10, 11, and 12 toothpicks.

After you have completed this exercise, check your answers to show that the sum of any two sides of a triangle is always larger than the third side.

Exercises

9.   A boy on a dock has a canoe tied at the end of a tight rope. The boy walks one metre along the dock towards *D* holding onto the rope. Does the canoe move 1 m, more than 1 m, or less than 1 m? Assume that *AC* is 12 m.

First draw a diagram to help you see the situation. In the diagram, what is the length of *CB*? How is *AB* + *BC* related to *AC*? Now answer the question.

3.3   *Right-angled triangles*

Thousand of years ago, the Babylonians used a property of right-angled triangles to help them with the construction of their buildings and other structures. Today that same property is used by architects, draftspersons and surveyors to help them in their work.

In a right-angled triangle, the longest side is the *hypotenuse*, and it is opposite the right angle. The hypotenuse in $\triangle ABC$ is side *AC*.

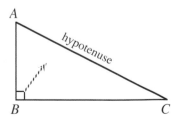

In the diagram, squares are drawn on the sides of a right-angled triangle. The sum of the areas of the squares drawn on the sides BC and AC is equal to the area of the square drawn on the hypotenuse, AB. This relationship among the areas of the three squares describes a basic property of right-angled triangles. Since c is the hypotenuse and a and b represent the lengths of the other two sides, we can write this relationship as

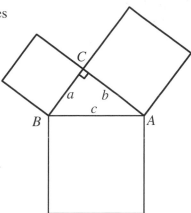

$$a^2 + b^2 = c^2 .$$

This property is named after the famous Greek mathematician, Pythagoras.

Pythagorean Theorem
The square on the hypotenuse of a right-angled triangle is equal to the sum of the squares on the other two sides.

$$a^2 + b^2 = c^2$$

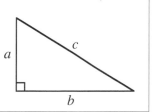

The Pythagorean Theorem is probably the most famous and most important result in geometry. It allows us to compute the unknown length in a right-angled triangle, if given the other two sides.

Example 1
Determine x in the diagram.

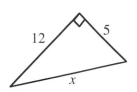

Solution
The length of the hypotenuse is x.
Using the Pythagorean Theorem,

$$x^2 = 5^2 + 12^2$$
$$= 25 + 144$$
$$= 169.$$
$$x = \sqrt{169}$$
$$x = 13.$$

Example 2

How far from the base of a vertical wall should the foot of a 6 m ladder be placed in order to reach 5 m up the wall?

Write your answer correct to one decimal place.

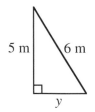

Solution

Let *y* represent the distance from the wall to the foot of the ladder, in metres.

In the diagram, the length of the ladder represents the hypotenuse of a right-angled triangle.

Use the Pythagorean Theorem to solve for *y*.

$$y^2 + 5^2 = 6^2$$
$$y^2 + 25 = 36$$
$$y^2 = 11$$
$$y = \sqrt{11}$$
$$y \doteq 3.3.$$

The ladder should be placed 3.3 m from the base of the wall.

Exercises

10.   Determine the value of the variable in each diagram.

    (a)

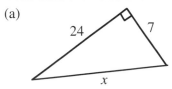

    (b)

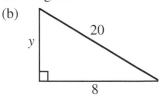

11.   How much farther is it to walk around the rectangular field to get from *A* to *B* rather than walking a straight line across the field from *A* to *B*?

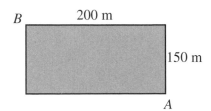

12. A pine log has a circular cross section with a diameter of 68 cm. What is the length of the sides of the largest square beam that can be cut from the log? Write your answer to the nearest centimetre.

3.4   *The circle*

The circle is an important shape in geometry and in the world around us. In order to understand the following examples you should be familiar with the fact that the sum of the angles at the centre of a circle is 360°.

Example 1

The smaller (minor) arc from A to B is $\frac{2}{5}$ of the circumference of the circle.

What is the measure of $\angle ACB$ at the centre of the circle?

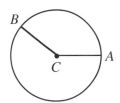

Solution

$$\angle ACB = \frac{2}{5} \times 360°$$
$$= 144°.$$

Example 2

Determine the number of degrees in the measure of the obtuse angle formed by the hands of a clock at 10:15 a.m.

Solution

At 10 a.m., the angle between the hands of the
clock is $\frac{2}{12} \times 360° = 60°$.

At 10:15 a.m. the minute hand has rotated $90°$.
Now consider the number of degrees the hour hand
has rotated.

In 60 minutes, the hour hand rotates $\frac{1}{12} \times 360° = 30°$.

In 15 minutes, the hour hand rotates

$$\frac{1}{4} \times 30° = 7.5°.$$

The measure of the obtuse angle formed by the hands
of the clock at 10:15 a.m. is $60° + 90° - 7.5° = 142.5°$.

Problems

1. Determine the number of degrees the minute hand of a clock rotates in the
   time interval from 9:15 a.m. to 9:35 a.m.

2. *ABC* is an isosceles triangle with $AB = AC$. *BD*
   is perpendicular to *AC*. Angle *A* is $40°$. What
   is the measure of $\angle DBC$?

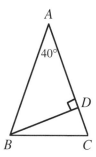

3. In the diagram, the measures of some
   angles are shown.
   What is the value of *x* ?

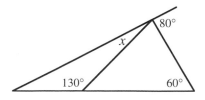

4. What is the measure of $\angle RPS$?

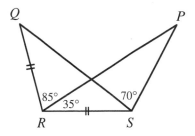

5. What is the measure of the angle through which the hour hand of a clock rotates between 9:20 a.m. and 11:50 a.m. on the same morning?

6. There are fifteen spokes equally spaced around a wheel. What is measure of the angle, in degrees, between two adjacent spokes?

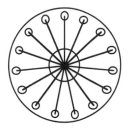

7. Determine the number of degrees in the measure of the obtuse angle formed by the hands of a clock at 11:18 a.m.

8. In the circle with centre O, the shaded sector represents 20% of the area of the circle. What is the size of angle AOB?

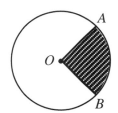

9. A regular pentagon has all sides and angles equal. If the shaded pentagon is enclosed by squares and triangles, as shown, what is the value of x?

10. In the diagram, the triangle shown is isosceles with $AB = AC$. What is the value of x?

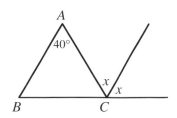

11. In triangle ABC, $AB = AC$, $BC = 12$ and $\angle A = 36°$. The line segment CD bisects angle ACB. What is the measure of CD?

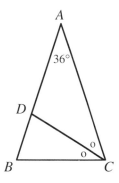

12. In the diagram, what is the value of $x + y$?

13. In the diagram, what are the values of a and b?

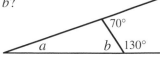

14. In the diagram, triangle PQT is equilateral and $QRST$ is a square. What is the measure of angle QRP?

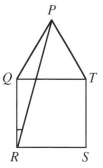

15. In the diagram, determine the value of x.

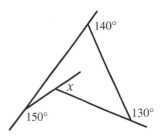

16. In $\triangle ABC$, BC bisects $\angle ABC$ and $AD = BD = BC$. Determine the measure of $\angle ACB$.

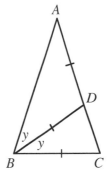

17. In quadrilateral $PQRS$, angle P equals $120°$ and angle Q is four times angle S. If angle R is $90°$, find the number of degrees in angle S.

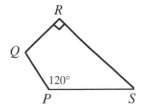

18. Determine the values of each of the variables in the diagram.

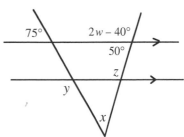

19. In the diagram, *ABCD* is a rectangle and *ABE* is an equilateral triangle. *M* is the midpoint of *BE*. Calculate $\angle BMC$.

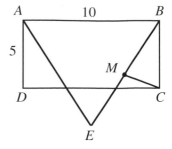

Chapter 4   Measurement

4.1   Perimeter

Many of the basic shapes and solids you see around you are made up of polygons. In this section we will discuss some of the properties of regular polygons and use them to help us calculate perimeter and area.

You will recall from the geometry chapter that polygons are named according to the number of sides. For example, a three-sided polygon is a triangle (remember that there are different types of triangles).

In the same way, any four-sided figure is a quadrilateral. We identify special types of quadrilaterals by their properties. Notice that every time one new property is added we have a new type of quadrilateral.

Name	Figure	Description
quadrilateral		A *quadrilateral* is a polygon with four sides.
trapezoid		A *trapezoid* is a quadrilateral with only two sides parallel.
parallelogram		A *parallelogram* is a quadrilateral with both pairs of opposite sides parallel.
rhombus		A *rhombus* is a parallelogram with all sides equal, that is, $AB = BC = CD = DA$.
rectangle		A *rectangle* is a parallelogram with one right-angle.
square		A *square* is a rectangle with all sides equal.

> The *perimeter* of any polygon is the sum of the lengths of all its sides.

Example 1

What is the perimeter of the figure *ABCDEF*?

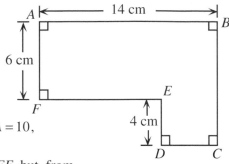

Solution

The perimeter is
$AB + BC + CD + DE + EF + FA$.
We are given $AB = 14$, $BC = DE + FA = 10$,
$DE = 4$, and $FA = 6$.
We don't know the length of CD and EF, but from
the diagram $CD + EF = AC = 14$.
The perimeter of figure *ABCDEF* is $14 + 10 + 14 + 4 + 6 = 48$ cm.

Here is another solution.
If we extend AF and CD to meet at G,
we form rectangle *AGCB*, as shown.
The perimeter of the given figure
ABCDEF is equal to the perimeter of
rectangle *AGCB*.
From the diagram, we note that
$FG = ED = 4$ and therefore
$AG = AF + FG = 10$.
Also $GC = AB = 14$.
The perimeter of rectangle *AGCB* is $14 + (6 + 4) + 14 + 10 = 48$ cm.

Exercises

1. In the diagram, all edges which
    meet form right angles. What is
    the perimeter of the figure?

    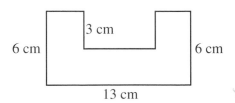

2. An isosceles triangle has two sides that measure 8 cm and 3 cm. What is
    the perimeter of the triangle?

3.   The perimeter of a rectangle is 96 cm. The width of the rectangle is
      15 cm. What is the length of the rectangle?

4.   Each of the equal sides of an isosceles triangle is 8 cm longer than the
      third side. The perimeter of the triangle is 49 cm. What are the lengths
      of the equal sides?

The Circle

The diagram illustrates some of the parts
associated with a circle.
AB is a diameter and *CD* is a radius.
Note that the radius *CD* is one-half the diameter *AB*.

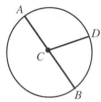

In a circle we use the term circumference rather than perimeter.
The *circumference* of a circle is calculated by using the formula

$$C = \pi d,\text{ where } d \text{ represents the diameter.}$$

Since the radius is one-half the diameter, then $d = 2r$ and the circumference of
the circle is

$$C = 2\pi r,\text{ where } r \text{ represents the radius.}$$

Note: It is impossible to determine an exact value for π. With the aid of
computers, π has been evaluated to millions of decimal places. In this
book we will write $\pi \doteq 3.14$.

Example 2

Determine the circumference of a circle with
diameter 8. Write the answer correct to one
decimal place.

8 cm

Solution

$C = \pi d$

$\doteq 3.14 \times 8$

$= 25.12.$

The circumference of the circle is 25.1 cm, correct to one decimal place.

Example 3
Determine the perimeter of the window in the diagram, given that arc *DEA* is a semicircle.

Solution
In order to calculate the perimeter of the window, we need to calculate $AB + BC + CD + $ arc *DEA*.

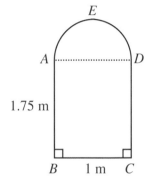

$$AB + BC + CD = 1.75 + 1 + 1.75$$
$$= 4.5.$$

Arc $AED = \frac{1}{2}$ (the circumference of the circle with diameter *AD*)

Since $AD = BC = 1$, then arc $AED \doteq \frac{1}{2} \times 3.14 \times 1$
$$= 1.57.$$

The perimeter of the window is $4.5 + 1.57 = 6.07$ m, or 6.1m, correct to one decimal place.

Exercise

5. Determine, to the nearest metre, the perimeter of the race track.

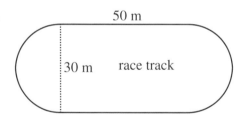

6. A bicycle wheel has a diameter of 50 cm. How far does the bicycle travel when the wheel makes 30 turns?

7. What happens to the circumference of a circle if the diameter is doubled?

4.2 *Area*

To find the area of a shape, you could choose a number of square centimetre tiles. You could then cover the shape with these tiles and, if they fit, count the number of tiles required. For example, if you wanted to determine the number of square metres of carpet required to cover a rectangular floor, you could cover the floor with pieces of paper, each one square metre, and then count the number of pieces. In practice there is an easier way, as you already know; you measure the length and width and multiply them. Area formulas allow us to replace the tedious covering and counting technique. Listed below are the area formulas you will need.

In the last section we started with the quadrilateral and ended our list of four-sided polygons with the square. Here it is easier to begin with the area of a square and work our way to the area of the quadrilateral, because the square is the simplest of the quadrilaterals to work with.

Square

$$\text{Area} = (a) \times (a)$$
$$\text{or} \quad A = a^2.$$

Rectangle

$$\text{Area} = (\text{length}) \times (\text{width})$$
$$= l \times w.$$
$$\text{or} \quad A = lw.$$

Parallelogram

To find the area of a parallelogram with base b and height h, cut off the triangle at the corner and slide it sideways to create a rectangle as shown in the diagrams.

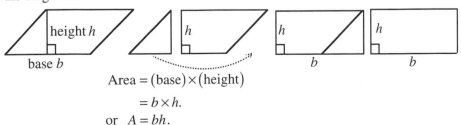

$$\text{Area} = (\text{base}) \times (\text{height})$$
$$= b \times h.$$
$$\text{or} \quad A = bh.$$

Since a rhombus is a parallelogram, we can use this same formula for its area.

Example 1

Determine the area of parallelogram *ABCD*.

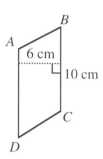

Solution

Note that the base is $BC = 10$ and the height is $h = 6$.

$$A = bh$$
$$= 10 \times 6$$
$$= 60.$$

The area is 60 cm^2.

Exercise

8. The area of a parallelogram is 48 cm^2. If the base is 12 cm, what is the height of the parallelogram?

9. Given a rectangle and a parallelogram with the dimensions shown, find the area of the shaded region.

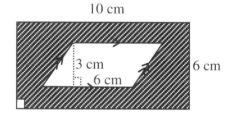

Trapezoid

To determine the formula for the area of a trapezoid, redraw the trapezoid, flip it, and attach as shown below.

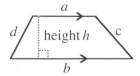
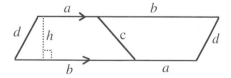

We now have a parallelogram as shown. The area of the parallelogram is

$$A = base \times height$$
$$= (a + b)h.$$

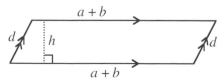

But this parallelogram has twice the area of the
original trapezoid. Therefore the area of the
trapezoid is one-half the area of this
parallelogram. That is, the area of the trapezoid
is one-half of the sum of the lengths of the
parallel sides multiplied by the height.

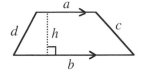

$$A = \frac{(a+b)}{2} \times (h)$$

$$\text{or} \quad A = \frac{h(a+b)}{2}.$$

Example 2
Determine the area of the trapezoid.

Solution

$$A = \frac{h(a+b)}{2}.$$

Since $a = 4$, $b = 12$, $h = 3$,

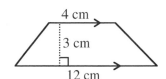

$$A = \frac{3(4+12)}{2}$$

$$= \frac{3(16)}{2}$$

$$= 24.$$

The area of the trapezoid is 24 cm^2.

Quadrilateral

There is no simple formula for the area of a quadrilateral.
(Sometimes we can divide the quadrilateral into other shapes whose areas we
can calculate.)

Area of a triangle.

Since the area of triangle *ABC* is one-half the
area of a parallelogram *ABCD*,

the area of $\triangle ABC = \frac{1}{2}bh$.

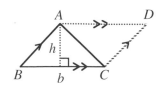

The height *h* of a triangle is the perpendicular
distance from a vertex to the opposite side.
Sometimes the opposite side must be extended
in order to draw the height. In $\triangle ABC$, *BC* is
the base and *AD* is the height.

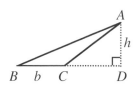

Note: There are three heights associated with any triangle.

Using *BC* as the base,
AD is the height.

Using *AC* as the base,
BE is the height.

Using *AB* as the base,
CF is height.

Exercises

10.   Are the three heights of a triangle equal to one another?

11.   What must be true about the sides of a triangle so that all three heights
       are the same?

12.   Determine the area of the trapezoid.

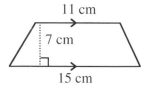

13.   In each row and column the dots are one cm
       apart. What is the area of the shaded region?

14. In the diagram, *ABCD* is a square with each side of length 4. If $AE = 1$ and $DF = 3$, what is the area of quadrilateral *AEFD*?

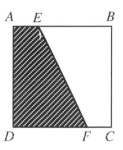

Heron's Formula

There is a remarkable formula known as *Heron's Formula* that expresses the area of a triangle in terms of the lengths of its three sides.

The formula is $A = \sqrt{s(s-a)(s-b)(s-c)}$,

where *a*, *b* and *c* are the length of the sides of the

triangle and *s* is one-half the perimeter, or $s = \frac{1}{2}(a+b+c)$.

Example 3
Determine the area of $\triangle ABC$, correct to one decimal place.

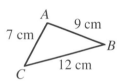

Solution
Since the lengths of the three sides are given, we can

use Heron's formula, that is, $A = \sqrt{s(s-a)(s-b)(s-c)}$.

For this example, $a = 12$, $b = 7$, $c = 9$, and $s = \frac{1}{2}(12+7+9) = 14$.

$$A = \sqrt{14(14-12)(14-7)(14-9)}$$
$$= \sqrt{980}$$
$$\doteq 31.30$$

The area of the triangle is 31.3 cm^2.

Example 4

Given that the area of $\triangle ADC = 45\,\text{cm}^2$, $AD = 5\,\text{cm}$ and $DB = 12\,\text{cm}$, determine the area of $\triangle ABC$.

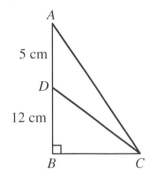

Solution

The area of $\triangle ABC = \frac{1}{2}(AB)(BC)$ where AB is the height and BC is the base. (Could BC represent the height and AB the base? Why?)

To find BC, we use the fact that the area of $\triangle ADC = 45\,\text{cm}^2$. But what will we use for base and height? If we look at the diagram sideways we see that we can use AD as base and BC as height.

The area of $\triangle ADC = \frac{1}{2}(AD)(BC)$

$$45 = \frac{1}{2}(5)(BC)$$
$$90 = (5)(BC)$$
$$BC = 18.$$

Now we can calculate the area of $\triangle ABC$.

The area of $\triangle ABC = \frac{1}{2}(AB)(BC)$, where $AB = 17$ and $BC = 18$.

$$= \frac{1}{2}(17)(18)$$
$$= 153.$$

Therefore the area of $\triangle ABC$ is $153\,\text{cm}^2$.

Activity

Alison claims to have discovered a way of finding the area of a circle by cutting it into twelve equal pie-shaped pieces and rearranging them into a shape resembling a parallelogram. Try this using the diagram below as a guide.

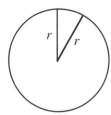 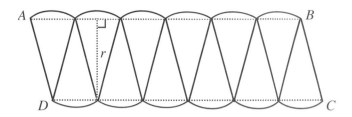

(i) What is the measure of the angle at the centre of the circle for each pie-shaped piece?

(ii) Alison says that dotted line *AB* is one-half the circumference of the circle. Do you agree?

(iii) Alison says that the area of the circle is approximately equal to the area of the parallelogram *ABCD*. Find the area of parallelogram *ABCD*. How does this compare with the actual area of the circle obtained by using $A = \pi r^2$?

(iv) If Alison did this by cutting the circle into 60 pieces, would the approximation of the area be better? (Why?)

Exercises

15. What is the area of a circle with radius 4 cm?

16. A square has an area of 36 cm^2. What is the perimeter of the square?

17. In the diagram, $BC = 12$ cm, $AB = 5$ cm, and the area of triangle *ADC* is 20 cm^2. Find the area of triangle *ABD*.

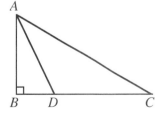

18. There is no formula for the area of a quadrilateral. Draw any quadrilateral and explain how you might find its area.

19. What is the area of quadrilateral *ABCD*?

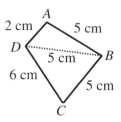

4.3   *Surface Area*

So far our discussion of area has been limited to two dimensional shapes. For a three dimensional shape, the sum of the area of its faces represents the surface area of the object. For example, to find the surface area of a three dimensional figure such as a cube, all you need to consider is the shape of each face. If you draw the net of a cube you can see all of its faces at once.

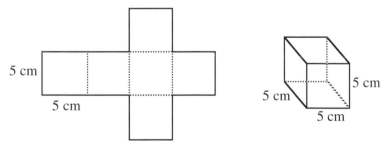

Each face is a square so the area of each face is $5 \times 5 = 25 \text{ cm}^2$.

The surface area of the cube is the sum of the areas of the six faces.

The surface area, *SA* , of the cube is $6 \times 25 = 150 \text{ cm}^2$.

Example 1
What is the surface area of the square-based pyramid shown?

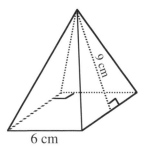

Solution

The square-based pyramid consists of five faces. The base is a 6 cm by 6 cm square. The four sides are identical triangles each with base 6 cm and height 9 cm.

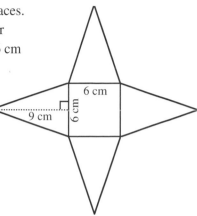

Area of the square base is $6 \times 6 = 36 \text{ cm}^2$.

Area of one triangle is $\frac{1}{2}(6)(9) = 27 \text{ cm}^2$.

Area of four triangles is $4 \times 27 = 108 \text{ cm}^2$.

The surface area of the pyramid is $36 + 108 = 144 \text{ cm}^2$.

Surface area formulas exist for some special surfaces. For example, the formula for the surface area of a sphere is $SA = 4\pi r^2$. An example of a sphere is a tennis ball.

Exercises

20. Calculate the surface area of a cube whose sides measure 6 cm.

21. Calculate the surface area of a box with length 12 cm, width 24 cm and height 7 cm.

22. What is the surface area of a ping-pong ball with radius 2 cm?

23. Calculate the surface area of the pyramid.

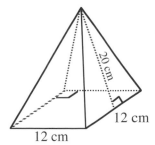

24.   The diagram shows a soup can with the top
      and bottom removed. The diameter of the can
      is 6 cm and the height is 15 cm. If the can is
      cut along the dotted line, what is the perimeter
      of the resulting rectangle? Try this on a soup
      can, but instead of the cutting the can, cut the
      label.

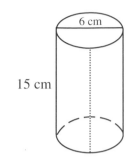

   (a)   What is the surface area of the can
         without the top and bottom?
   (b)   What is the surface area if the top and bottom are included?

25.   A circular piece of paper with radius 12 cm
      has the shaded sector removed, as shown. The
      remaining piece of paper is formed into a
      drinking cup in the shape of a cone. Try this
      yourself.

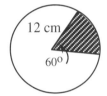

   (a)   What is the surface area of the curved surface of the cup?
   (b)   What is the circumference of the top of the cone?

4.4   *Volume*

Volume is the amount of space taken up by a three-dimensional object.

Volume is measured in cubic units, such as cubic centimetres $\left(cm^3\right)$, cubic

metres $\left(m^3\right)$, and so on. For example, if we want to find the amount of space

taken up by a box that measures 20 cm by 8 cm by 6 cm, we could fill the box
with 1 cm cubes and then count how many are required. In practice, volumes
are seldom determined by filling and counting. But we can use this concept to
find the volume of a regular solid without going through the actual process of
filling and counting.

A *regular solid* or *prism* is a three dimensional object in which the top and
bottom (or the ends) are the same shape and are parallel to one another. In
addition the sides are perpendicular to the top and bottom.

For example, the rectangular box at the right is a regular solid since the bottom (or the base) has the same shape as the top and both are rectangles with dimensions 20 cm by 8 cm.

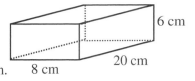

6 cm

20 cm

8 cm

To find the volume of a regular solid, such as this box, all we need to do is find the area of the base and then multiply by the height.

The volume of a regular solid (a prism) is (area of the base) x (height).

The area of the base is $8 \times 20 = 160 \text{ cm}^2$.

Since the height is 6 cm, the volume is $V = 160 \times 6 = 960 \text{ cm}^3$.

The three dimensional solid at the right is also a prism because the top and the base have the same shape, even though they are irregular. Note that the sides are parallel and perpendicular to the base. What is the volume of this solid, if the area of the base is 30 cm^2?

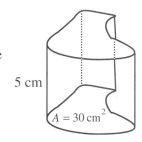

5 cm

$A = 30 \text{ cm}^2$

Exercises

26. A 5 cm by 5 cm by 5 cm cube is removed from a 10 cm by 10 cm by 10 cm cube. What is the volume of the remaining solid?

27. A solid rectangular block, measuring $4 \text{ cm} \times 6 \text{ cm} \times 8 \text{ cm}$, is made by using cubes with each side measuring 2 cm. How many cubes are necessary to make this block?

28. A cube has a volume of 125 cm^3. What is the area of one face of the cube?

29. Calculate the volume of the prism at the right.

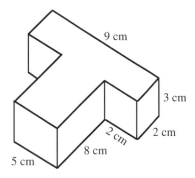

30. Is a cylinder a prism? (Why?)

31. Calculate the volume of the cylinder.

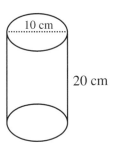

Problems

1. Three square tiles are placed as in the diagram. The total area is 300 cm^2. What is the perimeter of one of the squares?

2. In the diagram, *ABCD* is a trapezoid. What is the perimeter of the trapezoid?

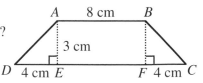

3. In the diagram, *PQ* is parallel to *RS*. The diameter, *PS*, of semicircle *PTS* is 12 cm. Determine the perimeter of the figure, correct to the nearest cm.

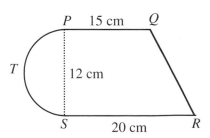

4. In the diagram, the line segment BC joins the centres of the two circles. If $BC = 8$ and $AC = 10$, what is the circumference of the smaller circle?

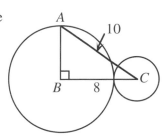

5. What is the greatest number of rectangular tickets, measuring 4 cm by 5 cm, which can be cut from a rectangular piece of cardboard measuring 25 cm by 28 cm?

6. What is the area of triangle PQR?

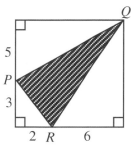

7. What is the area of the figure shown?

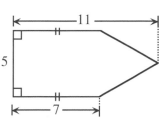

8. In the 9 by 7 rectangle shown, what is the area of the shaded region?

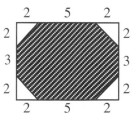

9. Four pipes, each of diameter 1 m, are held tightly together by a metal band as shown. How long is the band?

10. In the octagon shown, opposite sides are
    equal in length. If $AB + BC + CD = 14$ cm,
    $BC = DE$, $AB = 4BC$, and $CD = 2BC$,
    what is the perimeter of the octagon?

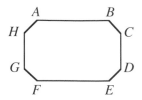

11. The largest possible circle is drawn inside a square. Then the largest
    possible square is drawn inside this circle. What is the area of the inner
    square as a fraction of the area of the outer square?

12. A rectangle has dimensions of 40 cm by 30 cm. A new rectangle is
    created by increasing the longer side by 20% and decreasing the shorter
    side by 20%. What is the difference in area between these two
    rectangles?

13. A circle of radius 12 cm is cut into two
    segments by a chord of length 12 cm. What is
    the perimeter of the shaded segment?

14. In the diagram, the dots are one unit apart
    horizontally and vertically. What is the area,
    in square units, of the figure shown?

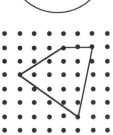

15. The area of the floor of a rectangular room is 195 m^2. One wall is a
    rectangle with area 120 m^2 and the other wall is a rectangle whose area
    is 104 m^2. If the dimensions of the room are all integers, what is the
    volume of the room?

16. A square with perimeter 20 is contained within a larger square of perimeter 28. What is the area of the shaded region?

17. The perimeter of square *Y* is 60 cm and the perimeter of square *P* is 32 cm. What is the perimeter of right-angled triangle *T*?

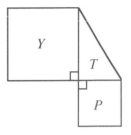

18. A circle is inscribed in trapezoid *PQRS*. If *PS* = *QR* = 25 cm, *PQ* = 18 cm and *SR* = 32 cm, what is the length of the diameter of the circle?

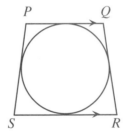

Chapter 5 Patterns

5.1 Keeping It Simple

Mathematics is often described as the science of patterns. We find and use patterns in many everyday events. A wallpaper pattern is a repeating design; a carpet pattern is symmetrical design; a dress pattern is a cut-out template; a model airplane pattern is a blueprint or a list of instructions; and a behaviour pattern is a predictable reaction to an event.

Let us begin by examining some patterns that use numbers, symbols and geometric shapes.

Exercise
1. What comes next?
    (a) 5, 11, 17, 23, 29, ...   (... means "*and so on*")
    (b) 3, 6, 12, 24, 48, 96, ...
    (c) 1, 2, 6, 24, 120, 720, ...
    (d) ⇦, ↖, ⇧, ↗, ⇨, ↘, ⇩, ...
    (e)

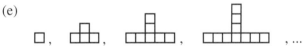

    (f) 5, 6, 8, 11, 15, 20, 26, 33, ...
    (g) ←O◣, ↙●◢, ↓O◥, ↘●▶, →O◣, ↗●◢, ...
    (h) 1, 1, 2, 3, 5, 8, 13, 21, 34, 55, ...

Finding a pattern can be an enjoyable activity and can also save a lot of work! We can often use pattern detection to find solutions to problems that may seem very time consuming or impossible. Consider the following problem.

Example 1
What is the units digit (or, as some call it, the ones digit) in the expansion of 2^{50}?

Comments
This doesn't look much like a pattern problem. In fact, it looks more like a very long calculation problem! But one of the most useful problem solving strategies is to try to *reduce a problem to a simpler one*, and try to find a pattern that may help solve the original problem.
(If you have a calculator, try to calculate 2^{50}. Could you use the calculator's answer to solve the problem?)

> ***Problem Solving Strategy***
> Reduce a problem to a simpler one, and try to find a pattern that will help solve the original problem.

Solution

Let's start with powers of 2 that we can evaluate easily.

$$2^1 = 2$$
$$2^2 = 2 \times 2 = 4$$
$$2^3 = 2 \times 2 \times 2 = 8$$
$$2^4 = 2 \times 2 \times 2 \times 2 = 16$$
$$2^5 = 32$$
$$2^6 = 64$$
$$2^7 = 128.$$

If we remember the original problem, we are only asked to find the *units* digit, so we should focus only on the last digit.

Power of 2	Exact Answer	Units Digit
2^1	2	2
2^2	4	4
2^3	8	8
2^4	16	6
2^5	32	2
2^6	64	4
2^7	128	8
2^8	256	6
2^9	512	2

Do you see the pattern in the units digit column?

The pattern 2, 4, 6, 8, repeats every four digits. All powers with a multiple of 4 in the exponent $\left(2^4, 2^8, 2^{12}, 2^{16}, ...\right)$ will have 6 as the last digit.

Now can we find the units digit of 2^{50}?

The closest multiple of 4 that is smaller than 50 is 48. So the units digit of 2^{48} must be 6. The units digit of 2^{49} must then be 2 and the units digit of 2^{50} must be 4.

Exercises

2. What is the units digit of
    (a) 2^{100}?   (b) 2^{75}?

3. Discover a pattern of powers of 3 to help find the units digit of
    (a) 3^{25} (b) 3^{75}.

4. What is the units digit of 9^{55}?

5. What is the units digit of 7^{100}?

6. Beads are placed on a string in the following pattern: red, green, blue, red, green, blue, red, green, blue, etc. What is the colour of the 40th bead on the string?

7. A pattern of shapes is repeated as shown below.

    What is the 211th figure in this pattern?

8. *Do you like to use computers?* Try using a computer to get an exact value of 2^{50} and check our solution in example 1.

9. The sequence 1, 1, 2, 3, 5, 8, 13, 21, 34, 55, ... that we saw at the start of this chapter was named after the mathematician and scientist Leonardo of Pisa, who was also called Fibonacci. Find out why this Fibonacci sequence is so famous and how it is connected to the *birth of baby rabbits*!

Pattern problems come in many different forms. Here is one that involves sums of numbers.

Example 2

What is the sum of $1+3+5+7+9+11+...+199$, (the first 100 odd positive integers)?

Solution

We can use the strategy of simplifying the problem and finding a pattern that we used with the other problems; but this time let's look at the *sums*.

Sum of first odd number: 1
Sum of first 2 odd numbers: $1+3=4$
Sum of first 3 odd number : $1 + 3 + 5 = \boxed{1 + 3} + 5 = 4 + 5 = 9$
Sum of the first 4 odd numbers: $1 + 3 + 5 + 7 = \boxed{1 + 3 + 5} + 7 = 9 + 7 = 16$
Sum of the first 5 odd numbers: $1 + 3 + 5 + 7 + 9 = \boxed{1 + 3 + 5 + 7} + 9 = 16 + 9$
$$= 25$$

Let's look for a pattern in our answers, 1, 4, 9, 16, 25.

Where have we seen a pattern of numbers such as this? These numbers are squares of the numbers 1, 2, 3, 4, and 5. So the sum of the odd numbers creates a surprising pattern of square numbers. For example,

$1+3+5+...+21=11^2$.

> The sum of the first n odd positive integers is n^2.

Example 3

What is the sum of the odd integers from 1 to 47, inclusive?

Solution

First, we must determine the number of odd integers between 1 and 47 inclusive. We can do this by adding 1 to 47 and dividing by 2. That is, there are $\dfrac{47+1}{2} = 24$ odd integers. (Can you explain why this works?)

Therefore the sum is 24^2 or 576.

Exercises

10.   What is the sum of the odd integers from 1 to 73 inclusive?

11. *What is the connection?* How does the diagram help us understand the pattern we found by summing odd numbers?

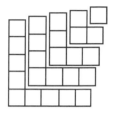

12. Use our strategy of simplifying the problem by finding a pattern of sums to predict the sum of $\frac{1}{2} + \frac{1}{4} + \frac{1}{8} + \frac{1}{16} + \frac{1}{32} + \frac{1}{64} + \frac{1}{128} + \frac{1}{256} + \frac{1}{512} + \frac{1}{1024}$.

13. What is the value of $50 - 49 + 48 - 47 + 46 - 45 + 44 - 43 + \cdots + 2 - 1$?

14. Look in your school or community library for books by Martin Gardner. He has written many books that will allow you to explore other interesting patterns.

5.2 *Sum of Consecutive Positive Integers*

One of the most famous stories about finding a pattern is associated with the accomplished mathematician, Carl Friedrich Gauss. In the school life of young Gauss, teachers often gave students complicated calculation problems . One day, so the story goes, a teacher gave the class the task of adding up the integers from 1 to 100; that is, finding the sum 1 + 2 + 3 + 4 + ... + 100.

Unfortunately for the teacher, Gauss, age 10, was a student in the class and after only a few moments gave the correct answer. Needless to say the teacher was surprised and challenged young Gauss to explain how he determined the answer so quickly. He explained that when he looked at the list of numbers to be added, a pattern became evident.

$$1 + 2 + 3 + 4 + 5 + ... + 96 + 97 + 98 + 99 + 100$$

Note that $1 + 100 = 101$, $2 + 99 = 101$, $3 + 98 = 101$, and so on.

Gauss wrote down the sum and underneath it wrote the same sum but in reverse order. He then added the two rows.

$$1 + 2 + 3 + 4 + ... + 97 + 98 + 99 + 100$$
$$100 + 99 + 98 + 97 + ... + 4 + 3 + 2 + 1$$
$$\overline{101 + 101 + 101 + 101 + ... + 101 + 101 + 101 + 101}$$

The sum of matching pairs is always 101. Since there are 100 numbers, the sum of $101+101+101+...+101$ is 100×101. But the pattern is used twice, and so the sum of $1+2+3+4+...+100$ is $\dfrac{100 \times 101}{2}$, which equals 5050.

Exercises

15. Use Gauss' method to find
     (a) the sum of $1 + 2 + 3 + 4 + 5 + 6 + 7 + 8 + 9 + 10 + 11 + 12 + 13 + 14 + 15 + 16$; then use a calculator to check your answer. (Which is the faster method?)
     (b) the sum of the first 50 consecutive positive integers.
     (c) the sum of the first 200 consecutive positive integers.

16. Does the same method work if you have an odd number of consecutive positive integers?
     (a) Find the sum of $1 + 2 + 3 + 4 + 5 + 6 + 7 + 8 + 9 + 10 + 11$, then use a calculator to check your answer.
     (b) Find the sum of the first 75 consecutive positive integers.

The work of Gauss can be extended to the sum of any list of consecutive positive integers as:

$$\frac{(number\ of\ consecutive\ positive\ integers)\ x\ (first\ number + last\ number)}{2}$$

Can you see the logic of this statement?

Did you know that …

Mathematicians use algebra to simplify the task of writing long instructions. Based on our work above, the sum of the positive integers from 1 to n (or in short form S_n) would be

$$S_n = \frac{n(n+1)}{2}.$$

Exercises

17. Who was Carl Gauss and what contributions did he make to mathematics?

18. What is the sum of the following consecutive integers?
    (a) $1+2+3+4+5+...+80$.
    (b) $1+2+3+4+...+125$.

19. What is the sum of $50+51+52+53+...+150$?

20. (a) Find the sum of $2+4+6+8+10+...+58+60$ by using Gauss' method.
    (b) Find the sum of $2+4+6+8+10+...+58+60$ by noticing that $2+4+6+8+10+...+58+60 = 2(1+2+3+4+...+30)$.

21. (a) Use Gauss' method to find the sum of the first thirty consecutive odd integers, namely, $1+3+5+7+9+...+55+57+59$.
    (b) How can you use the following pattern to get the same answer as in (a):
         $(1+2+3+4+...+58+59+60)-(2+4+6+...+56+58+60)$?

22. (a) What is the area of each triangular step figure shown below?

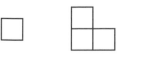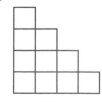

Fig. 1   Fig. 2   Fig. 3   Fig. 4

(b)   If the pattern continues, draw the 5th figure and find its area.

(c)   What will be the area of the 6th figure? 7th figure? 40th figure?

(d)   How is your work in finding the pattern for the areas of these figures related to Gauss's work in finding the sum of consecutive numbers?

What's the connection?

Do you see the connection between Gauss's method and the area of the triangular step figures?

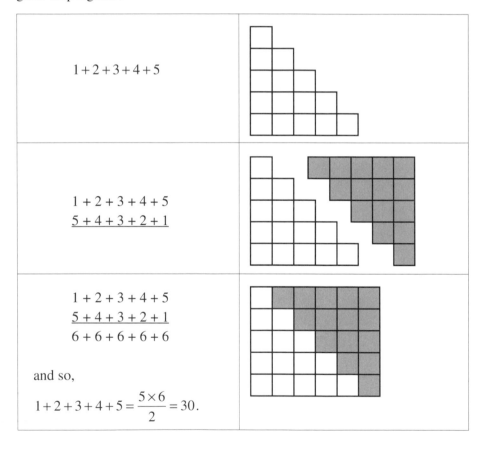

5.3   *Patterns and Algebra*

Algebraic expressions may be used to generate patterns of numbers.

For example, let us consider the value of the expression $4n + 1$ for $n = 1, 2, 3, 4, \ldots$.

> When $n = 1$, $4n + 1$ equals $4(1) + 1 = 5$.
> When $n = 2$, $4n + 1$ equals $4(2) + 1 = 9$.
> When $n = 3$, $4n + 1$ equals $4(3) + 1 = 13$.
> When $n = 4$, $4n + 1$ equals $4(4) + 1 = 17$.

The number pattern, or sequence, that this expression gives us is
$$5, 9, 13, 17, 21, 25, \ldots .$$
This is a pattern of numbers that begins with 5, with 4 added to each term to get the next term.

Did you know that …

To reduce the amount of writing, mathematicians use the symbol t_1 to mean the 'first term', the symbol t_2 to mean the 'second term', and so on.

So in the sequence 5, 9, 13, 17, we write $t_1 = 5, t_2 = 9, t_3 = 13, t_4 = 17$, and so on.
In fact, the nth term would be $t_n = 4n + 1$, since this was the expression we used to create the pattern.

Let's look at our original pattern one more time.

The expression $t_n = 4n + 1$ generates the numbers 5, 9 13, 17, 17, 21, … .
What clue is there in the expression 4n + 1 that the difference between successive terms is 4?
The number in front of the variable n (called the *coefficient* of n), is 4, and the difference between terms is 4. Will this be true in similar expressions?

Exercises

23. (a) Find the first six terms in the sequence defined by

$t_n = 5n - 2$, for $n = 1, 2, 3, 4, \dots$.

(b) What is the difference between successive terms in the sequence?

(c) How could you have predicted your answer in (b) before listing all the terms?

24. (a) Predict the difference between successive terms in the sequence defined by

$t_n = 8n + 3$, $n = 1, 2, 3, \dots$.

(b) Check your answer in (a) by listing the first six terms.

We can also create the algebraic expression from a pattern of numbers.

Example 1

What expression creates the number pattern: 5, 12, 19, 26, 33, 40, ... ?

Solution

When we examine the pattern, we see that 7 is added to to each term to get the next one. This tells us that 7 is the coefficient of n and the expression is of the form $7n + \boxed{?}$.

To complete the expression, we know that when $n = 1$, we must get 5. Since $7(1) + \boxed{?} = 5$, we must subtract 2 to make the equation true.

The expression must be $t_n = 7n - 2$, for $n = 1, 2, 3, \dots$.

Exercises

25. Check that $t_n = 7n - 2$, for $n = 1, 2, 3, \dots$, generates the pattern 5, 12, 19, 26, 33, 40, ...

26. Find an algebraic expression that generates each of these sequences of numbers.

(a) 1, 3, 5, 7, 9, 11, ...

(b) 1, 4, 7, 10, 13, 16, 19, ...

(c) 12, 17, 22, 27, 32, 37, ...

(d) 2, 9, 16, 23, 30, 37, 44, ...

(e) 1, 4, 9, 16, 25, 36, 49, 64, ...

27. Look at the pattern formed by the figures shown below.

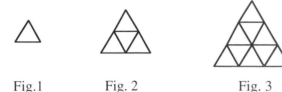

Fig.1   Fig. 2   Fig. 3

(a) Draw the 4th and 5th figures.
(b) Figure 3 is made from nine small triangles. How many small triangles are needed to make Figure 4? Figure 5?
(c) Without drawing, how many small triangles would be needed to make Figure 6? Figure 7? Figure 30? Figure 100?
(d) Write an expression that would indicate the number of small triangles needed to make the nth figure.
(e) How is this problem related to the work we did to find the sum of consecutive odd integers in example 22?

5.4 Pattern Explorations

Many problems allow you to extend and describe a pattern using the ideas you have learned and explored in this chapter.

Exploration 1 Triangular Patterns
Consider the chart below.

Row 1					1				
Row 2					2	2			
Row 3				3	4	3			
Row 4			4	6	6	4			
Row 5		5	8	9	8	5			
Row 6	6	10	12	12	10	6			
Row 7	7	12	15	16	15	12	7		

(a) Complete rows 8, 9 and 10.
(b) How *many* numbers would be in the 11th row? the 30th row?

Exploration 2   Pascal's Triangle.

This arrangement of numbers was named after the mathematician Blaise Pascal. It has many interesting patterns, and actually has many important uses in the study of probability.

Row 1							1						
Row 2						1		1					
Row 3					1		2		1				
Row 4				1		3		3		1			
Row 5			1		4		6		4		1		
Row 6		1		5		10		10		5		1	
Row 7	1		6		15		20		15		6		1

(a)   Complete the next three rows in Pascal's triangle.

(b)   How *many* numbers will be in row 12? in row 50? in row *n*?

(c)   Find the *sum* of the numbers in each row. What will be the sum of the numbers in row 12? in row 20? in row *n*?

Problems

1.   Given the products below, study the pattern of answers.

$$999 \times 222\,222 = 221\,999\,778$$
$$999 \times 333\,333 = 332\,999\,667$$
$$999 \times 444\,444 = 443\,999\,556.$$

Use the pattern to find $999 \times 888\,888$.

2.   How many squares are needed to make the 40th figure in this pattern?

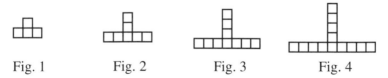

Fig. 1   Fig. 2   Fig. 3   Fig. 4

3.   Leslie buys five types of sports collector cards one at a time and always in the same order. He buys a hockey card first, then a baseball card, then a basketball card, then a football card, then a soccer card, and then starts again with a hockey card. What is the 37th card he buys?

4. Find the value of $490 - 491 + 492 - 493 + 494 - 495 + \ldots - 509 + 510$.

5. Beads are placed in the following sequence: 1 red, 1 green, 2 red, 2 green, 3 red, 3 green, with the number of each colour increasing by one every time a new group of beads is placed. How many of the first 100 beads are red?

6. Using only integers 1, 2, 3, 4, and 5, a sequence is created as follows: one 1, two 2's, three 3's, four 4's, five 5's, six 1's, seven 2's, and so on. The sequence appears as 1, 2, 2, 3, 3, 3, 4, 4, 4, 4, 5, 5, 5, 5, 5, 1, 1, 1, 1, 1, 1, 2, 2... .
   What is the 100th integer in the sequence?

7. Beads are placed in the following sequence: 1 red, 1 green, 3 red, 3 green, 5 red, 5 green, with the number of each colour increasing by two every time a new group of beads is placed. How many red beads are there in the first 100 beads?

8. The pattern of five figures is repeated as shown below.

   What is the 214th figure in the pattern?

9. The Ancients built square-based structures similar to those shown in the diagram. They began with 1000 identical cubes and wished to build as many structures as possible. The first structure contained two layers. Beside it was constructed a second structure with three layers. This process was continued as shown until the number of cubes left was not sufficient to build the next structure. How many cubes were left?

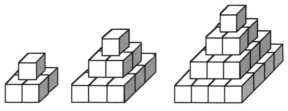

10. What is the sum of $5 + 10 + 15 + 20 + \ldots + 990 + 995 + 1000$?

11. What is the 100th number in the pattern 2, 5, 8, 11, 14, 17, ...?

12. A pattern begins 3, 7, 11, 15, 19, ... How many numbers will be in this pattern if the last term is 439?

13. What is the value of
$1 + 2 + 3 + 4 + 5 + 6 + ... + 13 + 14 + 15 + 14 + 13 + 12 + ... + 3 + 2 + 1$?

14. What is the value of $3 + 6 + 9 + 12 + 15 + ... + 99$?

15. The counting numbers are arranged in four columns as shown. If the pattern is continued, under which column will the number 101 appear?

P	Q	R	S
1	2	3	4
8	7	6	5
9	10	11	12
		14	13

16. (a) What number will be listed directly below 25 when this triangular array is continued in the same manner?

    1
    2   3
    4   5   6
    7   8   9   10
    ....

    (b) What number will be listed below 50? 75? 100?

17. (a) What is the units digit of $2^7 + 3^7$?

    (b) What is the unit's digit of $2^{34} + 3^{34}$?

Chapter 6   *Algebra*

6.1   *Understanding the Language*

When Carl Friedrich Gauss was a young student, he startled his teacher by mentally adding a long string of consecutive numbers in just a few seconds. How old was he when he did this? Well, if I told you that when you double his age and add 15 you would get 35, could you tell me his age?

Let's try another problem. I am thinking of a number. If I multiply it by 4 and then add 7, the result would be 24. What is the number?

This second problem is much harder than the first. In the first problem, you were probably able to determine that Gauss' age was 10 years, and you probably got the answer by logic or by making a thoughtful guess and then checking. For many of us, the second problem is more difficult because it is not as easy to guess and check – and this is where algebra becomes very useful.

We can use a symbol (a letter for example), which we call a variable, and proceed as though we know the number. Let the number be n; now multiply it by 4. But we can't multiply the number by 4 if we do not know the number! We can, however, write $4n$ to indicate the multiplication. (Notice that we do not write $4 \times n$; we just assume the multiplication sign). Next we can add 7 to $4n$ and set this equal to 24. Now we have:

$$4n + 7 = 24.$$

Since $17 + 7 = 24$, we know that $4n = 17$, so n must equal $4\frac{1}{4}$.

We got the answer by assuming we knew the number, representing it by a variable, and proceeding to do the required arithmetic. This is basically what algebra is all about. Algebra is a general form of arithmetic, in which we perform the usual operations we know but we use variables to represent unknown numbers.

Example 1

Select any positive integer and perform the following operations.

    1. Multiply the number by 4.

    2. Add 10.

    3. Divide by 2.

    4. Subtract 1.

    5. Divide by 2 again.

    6. Subtract the number that you first started with.

What do you get?

As an illustration, let's choose 8 as a starting number.

    Step 1. $4 \times 8 = 32$

    Step 2. $32 + 10 = 42$

    Step 3. $42 \div 2 = 21$

    Step 4. $21 - 1 = 20$

    Step 5. $20 \div 2 = 10$

    Step 6. $10 - 8 = 2$

The final answer is 2. This doesn't seem much like algebra, but if we examine the sequence more closely we'll see that we can use it. Will we get 2 if we start with 5? With 10? With 14?

Exercises

1.   Use the sequence of steps given above to answer each of the following.

    (a)   Repeat the same steps, but start with 10.

    (b)   Repeat the same steps again, but this time start with 17.

    (c)   Repeat the same steps again, but this time start with any number you choose.

    (d)   Look at your final answers. Do you notice any pattern?

Algebra can be used to show that we always get the same answer.

Let's follow the same sequence of steps but this time we will use the variable *n* to represent our number.

Choose a number	Example: Number is 10	Using Algebra: Number is n
1. Multiply by 4.	$10 \times 4 = 40$	$n \times 4 = 4n$
2. Add 10.	$40 + 10 = 50$	$4n + 10$
3. Divide by 2.	$50 \div 2 = 25$	$(4n + 10) \div 2 = 2n + 5$
4. Subtract 1.	$25 - 1 = 24$	$(2n + 5) - 1 = 2n + 4$
5. Divide by 2 again.	$24 \div 2 = 12$	$(2n + 4) \div 2 = n + 2$
6. Subtract the number that you first started with.	$12 - 10 = 2$	$(n + 2) - n = 2$
Final Answer	2	2

Aha! Algebra allows us to state that no matter what beginning number we select, if we perform these same six steps, our final answer will always be 2.

Example 2

Write an expression that is the same as $5(3n + 4)$.

Solution

Remember that algebra is just a general form of arithmetic. We know that $7 \times (8 + 9) = 7 \times 8 + 7 \times 9$. *Can you show why?*

So, $5(3n + 4) = 5 \times 3n + 5 \times 4$
$$= 15n + 20.$$

Exercises

2.   Write an expression that is the same as each of the following.

      (a) $4 \times (2n + 5)$ (b) $3 \times (4a + 7)$ (c) $(6n + 10) \div 2$

      (d) $(8a + 16) \div 8$

3. (a) Try different values with this sequence of operations to find a
         pattern to your answers. Explain the number trick by showing the
         steps using algebra.
         Pick a number.
         1. Multiply by 2.
         2. Add 5.
         3. Multiply by 3.
         4. Subtract 3.
         5. Divide by 6.
         6. Subtract your original number.

    (b) Try different values with this sequence of operations to find a
         pattern to your answers. Explain the number trick by showing the
         steps using algebra.
         Pick a number.
         1. Multiply by 5.
         2. Add 3.
         3. Multiply by 2.
         4. Add 4.
         5. Divide by 10.
         6. Subtract 1 from your answer.

4. Use what you know of algebra to create your own sequence of
    operations that always has a final answer of 5.

5. Find the value of $(p-1)(4p-2)$, when $p=3$.

6. Find the value of $\dfrac{(3a+b)(3a-b)}{a-b}$, if $a=4$ and $b=2$.

7. (a) How many circles will balance the triangle?

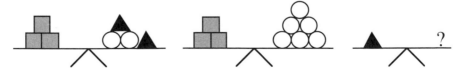

(b)   How many triangles will balance the square?

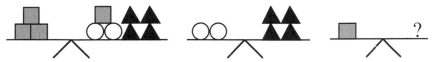

8.   A famous Arabian mathematician, Mohammed al-Khwarismi, wrote a
      book over one thousand years ago that used in its title the Arabic word
      al-jabr (meaning *broken bone*). The term algebra was derived from this
      word. Try to find out more about Mohammed al-Khwarismi and the
      early beginnings of algebra.

6.2   *Modelling with Algebra*

Let's use algebra to simplify some expressions.

Example 1
Find a simpler expression for each of the following.
(a) $5n + 9n + 2n$
(b) $5a + 9 + 6a - 5$

Solution
Using arithmetic, if you were asked what was "two 5's plus seven 5's" one
answer would be "nine 5's". This same rule works with algebra. We can add
together (or collect) algebraic terms that have the same symbol, or variable.
    (a) $5n + 9n + 2n = 16n$ (5 n's plus 9 n's plus 2 n's equals 16 n's).
    (b) $5a + 9 + 6a - 5 = 5a + 6a + 9 - 5$
                        $= 11a + 4$.
Why can't we add 11a + 4 to get 15a?

We can also use algebra to help us model a problem and to represent the steps
in the solution.

Example 2
The sum of four consecutive odd integers is 712. What are the integers?

Comments

One way to do the problem might be for us to make a series of guesses that would, hopefully, lead us to the correct answer.

We could, for example, try $23 + 25 + 27 + 29$, but this sums to 104, which is not even close.

We might guess higher and try $201 + 203 + 205 + 207$, but this sums to 816, which is too high.

Let's put some thought into our guesswork. We were told that four consecutive odd integers sum to 712, so we might try dividing 712 by 4 which gives us 178. This isn't odd, but it should be useful in making a good guess. Can you now determine the numbers?

Knowing a little algebra allows us to eliminate guesswork.

Solution

Let's use the variable n to represent the smallest number in our list of four. If n is the smallest number, what is the next odd number? Well, we have to add 2 to any odd number to get to the next odd number. So if n is the smallest, then $n+2$, $n+4$ and $n+6$ are the other three.

How can we use these to solve the problem? We know that the four numbers must add to 712, and so

$$n + n + 2 + n + 4 + n + 6 = 712.$$

Now we simplify the equation. We can 'collect' the four n's and add up the numbers, giving us

$$4n + 12 = 712.$$

The equation is now in a form that we can solve.

$$4n + 12 = 712$$ *What must 4n be equal to?*
$$4n = 700$$ *What number multiplied by 4*
$$n = 175.$$ *gives 700?*

This tells us that the smallest odd number is 175 and so the four consecutive odd numbers are 175, 177, 179, and 181. (Do they add to 712?)

Exercises

9. Write a simpler expression for each of the following.
   (a) $3x + 5x + 4x$
   (b) $7m + 5m - 2m$
   (c) $4t + 7 + t + 6$
   (d) $a + 3a + 4 + 7a + 5 - 2a$
   (e) $m + 3 + m + 6 + m + 4 + m + 9$

10. Solve each equation to find the value of n.
    (a) $3n + 5 = 29$
    (b) $6n - 4 = 50$
    (c) $n + n + n + n + n = 100$
    (d) $n + n + 1 + n + 2 = 48$

11. A certain number is multiplied by 8 and then increased by 4 to give the answer 100. Find the original number.

12. A certain number is multiplied by 7 and then decreased by 6 to give the answer 85. Find the original number.

13. The sum of three consecutive odd integers is 57. What are the integers?

14. The sum of five consecutive integers is 190. What are the integers?

15. The sum of four consecutive even integers is 1268. What are the integers?

16. Try to solve these equations.
    (a) $5n + 28 = 6n$ (b) $7y - 9 = 2y + 46$

6.3   *Using Our Algebra Skills*

As we become more familiar with algebra, we can use our new skills to solve some interesting problems and puzzles.

Example 1

There are 72 grade 8 students at the school. There are 6 more boys than girls in grade 8. How many girls are in grade 8?

Solution

With most problems, there are many ways that we can begin.

Let's say that we use g to represent the number of girls.

How could we represent the number of boys?

Well, we could say b represents the number of boys.

We know that there are 72 students in grade 8, so $b + g = 72$.

But the number of boys, b, is 6 more than the number of girls, g. This tells us that $b = g + 6$.

Using the fact that $b = g + 6$, we can then say

$$g + 6 + g = 72$$
$$2g + 6 = 72 \quad \text{\textit{We now have an equation we can solve!}}$$
$$2g = 66$$
$$g = 33.$$

There must be 33 girls and 39 boys in grade 8.

Example 2

A store has bicycles and tricycles for sale. How many of each are for sale if you count 99 wheels and 42 frames in the store?

Solution

Let's represent the number of bicycles by b and the number of tricycles by t.

We know that there are 42 frames in the store, so the total number of bicycles and tricycles is 42, or

$$b + t = 42.$$

Each bicycle has 2 wheels and each tricycles has 3 wheels, so

$$2b + 3t = 99.$$

Now what? There are two ways we could proceed.

Method 1

If you are unsure of the algebra, you can solve it by guess and check. The table below might help organize your work.

	b	t	$2b + 3t$
$b + t = 42$	20	22	$2b + 3t = 2(20) + 3(22) = 106$

Our guess of $b = 20$ and $t = 22$ proved to be incorrect. We can extend the table to make more guesses, but what do we know that might reduce the guesswork? Well, no matter what we choose for the value of b, the value of $2b$ must be even (why?). But $2b + 3t = 99$, and if $2b$ is always an even number, then $3t$ must be odd (Why?). If $3t$ is odd, then t must be odd. By golly, you were wasting time trying $b = 20$ and $t = 22$. They should both be odd numbers. Now complete the table using odd numbers for b and t.

b	t	$2b + 3t$
19	23	$2(19) + 3(23) = 107$
21	21	$2(21) + 3(21) = 105$
23	19	$2(23) + 3(19) = 103$
25	17	$2(25) + 3(17) = 101$
27	15	$2(27) + 3(15) = 99$

There are 27 bicycles and 15 tricycles in the store.

Method 2

Let's see how algebra helps us avoid the guesswork.

If $b + t = 42$ this means that $b = 42 - t$.
(For example, in arithmetic if $7 + 5 = 12$, then we can also say $7 = 12 - 5$).
So in the equation $2b + 3t = 99$, we can replace b with $42 - t$.

$$2(42 - t) + 3t = 99$$
$$84 - 2t + 3t = 99$$
$$84 + t = 99$$
$$t = 15.$$

There are 15 tricycles and $42 - 15$ or 27 bicycles in the store.

Exercises

17. Petra delivers papers to 80 homes, some of which are houses and the rest are apartments. If there are 16 more apartments than houses on Petra's paper route, to how many houses does Petra deliver papers?

18. A pet store sells only birds and puppies. Tom works at the store and counts the animals. He counts 30 heads and 94 feet. How many birds are in the store?

Problems

1. Pat was paid $150 for 5 days work. Each day after the first, she was paid $5 more than on the previous day. What was she paid on the last day?

2. A bicycle and a stand costs $330. The same bicycle with a stand and a light costs $370. The cost of the light is four times the cost of the stand. Find the cost of the bicycle alone.

3. Jen doubled a number, then added 6 to get the answer 28. Had she first added 6 to the original number, then doubled this, what would have been her answer?

4. If three times a number is 36, what is four times the number?

5. A burger, an order of fries and a soft drink cost $2.80. A burger and an order of fries cost $2.05. Two burgers and one order of fries cost $3.30. How much will a burger, two orders of fries and a drink cost?

6. Each of the letters *D*, *N*, *W*, and *P* in the chart represents a different number. The sums of the rows and columns are given. What is the sum, *T*, in the first column?

				Row Sum
D	D	D	D	28
D	D	N	N	30
N	W	P	D	20
P	P	W	N	16

Column Sum | *T* 19 20 30

7. If *n* is a positive integer, which, if any, of the following would always be divisible by 2? Explain why.

    (a) $n(n+2)$

    (b) $n(n+1)$

    (c) $n+n$

    (d) n^2

8. Pick any number. Add 5. Multiply by 2. Subtract 4. Divide by 2. Subtract the number you started with. What is your answer? Use algebra to explain the pattern of answers.

9. Basketball players score two points for a regular basket and three points for a long shot. Janine scored 36 points in a game. How many two-point and three-point baskets could Janine have made in the game? Find all possible answers.

10. In a group of 60 students, there are $\frac{2}{3}$ as many boys as girls. How many boys are there?

11. A man's estate valued at $120 000 is left in a will to his widow, his son, and his two grandchildren, as follows:
    To each of two grandchildren a certain sum, to the son four times as much as to the two grandchildren together, and to the widow $20 000 more than to the son and grandchildren together. How much goes to each?

12. Which of the following five expressions is always an odd integer, for any integer *n*? Explain why.

    (a) $2n$ (b) n^2 (c) $\sqrt{n}$ (d) $2n+1$ (e) $\frac{n}{2}$

13. In a magic square, the sum of each row, column, and diagonal is the same. In the magic square to the right, what are *A*, *B* and *C*?

16	2	A
C	10	E
B	D	4

14. When half a number is increased by 10, the result is 60. What is the original number?

15. If Δ is an operation such that $a \Delta b = ab + \frac{a}{b}$, what does $6 \Delta 2$ equal?

16. Fran purchased 150 marbles, some red, some white, and some blue. If she purchased twice as many white marbles as red marbles, and three times as many blue marbles as red marbles, how many red marbles did she purchase?

17. The Cycle Shop received a shipment of unassembled tricycles and bicycles. There were 20 frames and 48 wheels in the shipment. How many bicycles were in the shipment?

18. If $a \Delta b = \frac{a+b}{4}$, what is the value of $(7 \Delta 5) \Delta 1$?

19. The diagram shows a magic square in which the sums of the numbers in any row, column or diagonal are equal. What is the value of n?

8		
9		5
4	n	

20. Juan and Mary play a two-person game in which the winner gains 2 points and the loser loses 1 point. If Juan won exactly 3 games and Mary had a final score of 5 points, how many games did they play?

21. In the 4×4 square shown, each row, column and diagonal must contain each of the numbers 1, 2, 3, and 4. Find the value of K and the value of N.

I	F	G	H
T	2	J	K
L	M	3	N
P	Q	1	R

Chapter 7   *Counting and Probability*

7.1   *Counting and Averages*

We all know how to count. Even if asked to count by 2's, or by 3's, we can all manage a lengthy count. So why, then, is there a whole chapter in this book devoted to counting? Well, the truth is that applications of counting can get very tricky, so it's worth considering some of the ways in which simple counting can help us to solve quite difficult problems.

Example 1
A box contains five marbles, one each coloured red, green, blue, white, and yellow. If a girl reaches into the box and withdraws one marble, what are the chances that it is red?

Solution
You are probably able to say pretty quickly that she has one chance in five that the marble is red.

This gives us an answer, but what do we mean by one chance in five? And how does this relate to counting? The second question is easy to answer. If we can't count we'd never know how many marbles there are in the bag, so we wouldn't be able to say one chance in *five*. The first question is more challenging, and we'll just let it sit for a while. We'll get back to it soon.

You already know how to add integers manually, and recall that we developed a formula in an earlier chapter for finding the sum of consecutive integers. You also know that from a set of numbers we can determine various averages. In this book we are concerned only with the arithmetic average (or mean), which we determine by dividing the sum of the numbers by the number of elements in the set. Let's look at a couple of applications of this average.

Example 2
Betty plays golf. Her average score for seven games is 84. She then plays an eighth game, scoring 76. What is her average for the eight games?

Solution

To determine her average we need the sum of her scores for the eight games. While we do not know her score for any of the first seven games, we can say that if her average was 84, then the sum of the seven scores was 84×7, or 588. Then the sum of her scores for the eight games is $588 + 76$, or 664. Now we have the required sum so, dividing the total number of games by 8, we determine that her average score is $664 \div 8 = 83$.

Exercises

1. In a set of ten numbers, the average of the first four numbers is 12, and the average of last six numbers is 17. What is the average of all ten numbers?

2. Joe is a bowler. His average for ten games was 156.5. He scored only 116 on his tenth game. What was his average for the first nine games?

3. (a) Find the average of the numbers 10, 11, 12, 14, 16, 17, 18, and 22.
   (b) Add the number 105 to the set and determine the average of the nine numbers. Does this suggest that the term average may not give quite what we think it does?

7.2   *How Many Numbers?*

One of the interesting questions we meet is that of how many numbers can be created under certain conditions. For example, if we are given the digits 4, 5 and 6, how many different three- digit numbers can we make? In this example, and in all others in this section, we allow any given digit to be used only once in each number.

We can deal with this by simply writing all possible numbers: 456, 465, 546, 564, 645, and 654. There are six possible three-digit numbers.

This method isn't very practical if we are given the digits 1, 2, 3, 4, 5, 6, and 7, and asked how many possible three-digit numbers there are, because it would take too long to write them all down. There must be a short way of determining how many numbers there are, and we can see what that method is by considering a different example.

Example 1

A teacher has four different books and wishes to give one book to each of Dorcas and Jessie. In how many different ways can she give each girl a book?

Solution

If we label the four books A, B, C, and D, then Dorcas can receive any of the four. If Dorcas receives book A, Jessie can receive any one of B, C, or D. That is, for Dorcas receiving book A, there are three choices of books for Jessie. But it is also true that if Dorcas receives book B, there are three books that can be given to Jessie; A, C, or D. In fact, for any choice given to Dorcas, there are three choices from which to pick one for Jessie.

There are, then, $4 \times 3 = 12$ ways that the teacher can choose two books to give the girls. Note that this includes the possibility that Dorcas receives book B and Jessie receives book C, and also that Dorcas receives book C and Jessie book B.

From this example we can identify the basic idea of assigning objects (all of which are different):

> *If we can take some action in m different ways, and for each of these we can take a second action in n different ways, then there are $m \times n$ ways of taking the two actions.*

It is now easy to do the problem posed earlier. If we are given the digits 1, 2, 3, 4, 5, 6, and 7 (only one of each), and asked how many different three-digit numbers there are, we need only note that there are seven choices for the first digit (any of the seven given), six choices for the second digit (any of those given except the one chosen for the first digit), and five choices for the third digit. Hence there are $7 \times 6 \times 5 = 210$ possible three-digit numbers.

Here's an interesting extension of this problem. How many of these 210 numbers are even, and how many odd? Well, if we want a number to be even, it's the ones digit that's important; it has to be even. In this case, the ones digit must be 2, or 4, or 6, so there are three choices for the ones digit. Now there are only six choices for the first digit, and then five choices for the second digit.

Hence there are $6 \times 5 \times 3 = 90$ even numbers, and then there must be $210 - 90 = 120$ odd numbers.

Exercises

4.   Given four different coloured hats and two people, in how many ways can the two people be given a hat?

5.   How many trains consisting of one engine, one mail car, and one passenger car can be made up if there are two different engines, five different mail cars, and four different passenger cars?

6.   If you are given one each of the digits 2, 3, 4, 5, and 6, how many different three-digit odd numbers can be created?

7.   Given the same set of digits as in exercise 6, how many three-digit numbers can be created having a first digit of 5?

7.3   *Probability*

Probability plays an important role in our lives. We might hear on the radio that the probability of rain today is 30%, or that the probability of winning a game is 0.3. Closely associated with the term probability is the term *odds*; we might hear that the odds of winning a game are 3 to 8. This means simply that in 11 plays of the game we expect to win 3, so we can say that our probability of winning is $\frac{3}{11}$ and our probability of losing is $\frac{8}{11}$.

Calculating the probability of natural events such as weather, accidents, medical diseases, and so on, is an extremely complicated task, and we will make no effort to attempt such things in this book. However, it is easy to calculate the probability of an event if we can easily count the number of possible happenings and the number of these that contain some particular property.

Example 1

Using each of the digits 4, 5, 6, 7, and 8 only once, create a three-digit number. What is the probability that it has only even digits?

Solution

We first determine how many three-digit numbers there are. Since we have five digits, there are 5 choices for the first digit, 4 choices for the second, and 3 choices for the third. There are $5 \times 4 \times 3 = 60$ possible three-digit numbers.

Now we determine how many three-digit numbers have only even digits. Here we can use only 4, 6 and 8, so there are 3 choices for the first digit, 2 choices for the second, and 1 choice for the third. There are $3 \times 2 \times 1 = 6$ possible three-digit numbers that have only even digits.

Now we can say that 6 of the 60 possible three-digit numbers satisfy the requirement that all digits are even, so we can now answer the question. The probability that all three digits are even is $\frac{6}{60} = \frac{1}{10}$.

In counting outcomes we must be absolutely certain that we can distinguish all possible outcomes. This can be difficult. For example, in problems in which we discuss the outcomes of rolling two dice, it is easy to see that there is only one way to roll a pair of 2's, since both dice must show 2. However, there are two ways to roll a 3 and a 1, and we can't tell the difference if the dice are the same. If we use one red die and one green die this difficulty disappears because we can see the difference between a red 3, green 1 and a red 1, green 3. From this we see that we can help our thinking by insisting that things that might be alike be coloured differently.

Example 2
If two dice are rolled, what is the probability that the sum of the spots showing is 9?

Solution
Let one die be red, the other green.
There are 6 possible numbers on the red die and 6 possible numbers on the green die.
There are $6 \times 6 = 36$ possible combinations of the dice.
To obtain a sum of 9 we list the possibilities in the following table.

Red die	6	5	4	3
Green die	3	4	5	6

There are 4 ways of having a sum of 9.
The probability that the sum is 9 is $\frac{4}{36} = \frac{1}{9}$.

Exercises

8.   What is the probability of rolling a sum of 11 with two dice? What is
      the probability of getting any sum other than 11?

9.   You are rolling two 6-sided dice, but one die is numbered from 3 to 8
      and the other die is numbered from 4 to 9. What is the probability of
      rolling a sum of 14 with these two dice?

10.   A jar contains 4 red candies, 6 orange candies, 3 purple candies, and 7
      yellow candies. If you select one candy from the jar without looking,
      what is the probability that you will select an orange candy? What is the
      probability that you will select either a yellow or red candy?

11.   Given one each of 3, 4, 5, 6, and 7, you are asked to create a two-digit
      number. What is the probability that it is odd? (Think about the ones
      digit first).

7.4   *How Many Ways Revisited*

In all examples considered so far it has been stressed that there is only one of
any object considered. There are, however, all kinds of situations where
things are repeated: 343 has two 3's, we might give Mary four red books, and
so on. How does this change what happens when we count the number of
ways something can be done? Well, it makes things more difficult, and we
have to be much more careful in counting. Let's consider a problem.

Example 1
Tanya has the single digits 3, 3, 4, 4, and 5. Using these, how many different
three-digit numbers can she create?

Solution
There are three cases. Tanya can use two 3's and one other digit, or two 4's
and one other, or one of each of the different digits. (Can you see that the first
two cases will have the same number of possibilities?)

If she uses two 3's, she must use either one 4 or one 5. The numbers she can create are 334, 343, 433, 335, 353, 533. If she uses two 4's, she must use either one 3 or one 5. The numbers she can create are 443, 434, 344, 445, 454, 544. In each case there are six possible numbers. If she uses one 3, one 4, and one 5, she can create $3 \times 2 \times 1 = 6$ numbers. In total she can create $6 + 6 + 6 = 18$ numbers.

We can approach problems involving more than one of some of the objects by carefully considering cases, then by listing possibilities where necessary, and using what we learned earlier when possible.

Exercises
12. Given two 7's, one 4 and one 9,
    (a)   how many two-digit numbers can be formed?
    (b)   how many three-digit numbers can be formed?

13. Given three 6's, two 3's and one 8, how many three-digit numbers can be formed?

14. A chocolate bar dispensing machine contains three different kinds of chocolate bars, and has a lot of each kind. List the number of ways you can choose a selection of three bars, if the order in which you select doesn't count?

Be Careful

We must be particularly cautious in calculating the probability of things happening when there are objects that are alike. This is because there isn't always the same number of ways to determine all possible outcomes.

Example 2
Given two 5's and two 6's, calculate the probability of randomly drawing two numbers from a box to form
                  (a) the two-digit number 55, and
                  (b) the number 56.

Solution

Using two 5's and two 6's, the only numbers we can form are 55, 56, 65, and 66. It looks as though the probability of any number is $\frac{1}{4}$. But wait! If all the numbers were different, we'd be able to form $4 \times 3 = 12$ numbers, and now we can only create 4. What's changed?

Well, if we think about the two 5's as having different colours, say one yellow and one red, and the two 6's as also having colours, say one white and one blue, then we can form 55 in two ways, using the yellow one first then the red one, or using the red one then the yellow. There are *two* ways of making 55.

Now if we think of 56, we can use the yellow 5 with either the white 6 or the blue 6, and we can also use the red 5 with either the white 6 or the blue 6. There are *four* ways to make 56.

With the same reasoning, there are four ways to make 65 and only two ways to make 66. Aha! There are still 12 numbers possible, but a lot of them are the same. What this means, of course, is that the probability of creating 55 is $\frac{2}{12} = \frac{1}{6}$, and the probability of creating 56 is $\frac{4}{12} = \frac{1}{3}$.

From this example you can see the difficulty in trying to do this kind of problem. We are not going to pose any probability questions using repeated objects, and we advise you to be very, very careful if you are asked to do any.

Problems

1. The average of four numbers is 5. Two of the numbers are 3 and 4. What is the sum of the other two numbers?

2. The sum of three numbers is 180. The first number is 10 more than the average of the three numbers and the second number is 4 less than the average. What is the third number?

3. Lynn writes the numbers 26, 28, 29, 38, and 69, and realizes that one of the numbers is the average of the other four. What is that number?

4. A farmer grows pumpkins on her farm. Each pumpkin plant yields 8 pumpkins. She has 12 hectares of pumpkins and she harvests 6912 pumpkins. On average, how many pumpkin plants does she have on each hectare?

5. Jean writes five tests and achieves the marks on the graph. What is her average mark on these five tests?

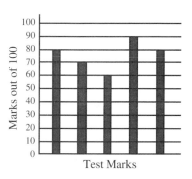

Test Marks

6. A bag contains 5 yellow marbles, 10 red marbles and 15 white marbles. If one marble is selected at random, what is the probability that it is yellow?

7. What is the average (mean) of the consecutive numbers
    (a)   10, 11, 12, 13, 14, 15, 16, 17, 18, 19, 20
    (b)   50, 51, 52, 53, 54, 55, ..., 95, 96, 97, 98, 99, 100
    (c)   10, 11, 12, ..., 19, 20, 21?
    Can you think of a quick way to find averages of consecutive numbers?

8. What seven consecutive numbers have an average of 20?

9. In a set of 15 numbers, the first 5 numbers have an average of 22, and the last 10 numbers have an average of 30. What is the average of all 15 numbers?

10. The average of five numbers is 12. When a sixth number is included, the average is 14. What is the sixth number?

11. Gainesville football team scored a total of 97 points in their first four games. After the fifth game, the team had averaged 27 points per game. How many points did Gainesville score in their fifth game?

12. Joan scored an average of 12.5 points in her first four basketball games, 15 points in her next five basketball games, and scored 20 points in her final basketball game. What was Joan's average points scored over all ten basketball games?

13. Using the letters of MATH, how many different three-letter arrangements can be formed? You may use each letter only once in each three-letter combination. Note: The combinations ATM, TMA and AMT are all different arrangements.

14. The pages of a book are numbered from 1 to 300 inclusive. How many digits are needed to number the pages of the book?

15. The five members of a curling team are getting their picture taken. If they are asked to line up in one straight line in front of the photographer, how many different arrangements are possible?

16. Of 35 students in a math class, 24 own roller blades and 19 own a skateboard. Two people do not own either. How many people own both a skateboard and roller blades?

17. An integer is chosen at random from 1 to 60 inclusive. What is the probability the integer chosen contains the digit 4?

18. Bob has six cards. He has two 5's, two 6's, one 7, and one 8. How many 3-digit *odd* numbers can be formed from the cards?

19. Pierre has two sets of twelve cards each numbered 1 to 12. One set of cards is red and the other is blue. If one card is selected from each deck, what is the probability of the two cards summing to 15?

20. (a) If a and b are positive integers, how many ordered pairs (a, b) are there such that $a + b = 25$? [One pair is $(4, 21)$ and a second one is $(21, 4)$].

     (b) Can you repeat this problem for a, b and c all positive integers and with $a + b + c = 26$? Now you are counting the number of ordered triples (a, b, c).

Chapter 8   Miscellaneous Problems

In the first seven chapters of this book, you reviewed a number of things and were introduced to several new mathematics topics. Most of the exercises and problems in each section related directly to the discussion in that chapter. Now that you have experienced a number of approaches to solving problems, you should be more confident in tackling problems which require creativity or unusual thought.

In this chapter you will be exposed to problems that are interesting and require a variety of mathematical topics in their solution. If at times you come to a dead end in your solution, don't give up. Take your time and think about the problem or discuss it with your friends before referring to the solution. Don't forget that to be a successful problem solver you need time for the creative juices to flow, you require perseverance, and above all you must be prepared to do some hard work. Enjoy yourself as you work your way through the last set of problems.

Example 1

The diagram is a map of roads linking the two villages of Amy's Cove to Burley, passing through Hillsdale, Fractionville, Deepwell, Manhill, and Steamroller.
The number on each section of road is the length, in kilometres, of that road. (The roads are not necessarily straight.)
How long is the shortest route from Amy's Cove to Burley?

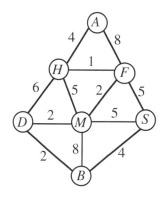

Solution

In order to solve this problem we examine various routes from *A* to *B* and record the distances. We will do this by listing the different routes and recording the shortest distance to a town by writing that distance in a box beside each letter.

From *A* the shortest route to *H* is 4 km. Place ☐4 beside *H*.

From *A* there are two routes to *F;* one from *A* to *H* to *F* which is 4 + 1 = 5 km, and the other directly to *F* which is 8 km. Write 5 beside *F* to indicate the shortest distance to that point.

Now, do we go from *H* to *D*, *H* to *M*, *F* to *M*, or *F* to *S?* Well, common sense says that we should choose the next step so that the total distance from *A* is as small as possible. (And there is no mathematical rule to help us.) Let's consider the possible choices, noting that the number beside *H* or *F* gives the distance from *A*.

Route	Distance
A to *H* and *H* to *D*	4 + 6 = 10
A to *H* and *H* to *M*	4 + 5 = 9
A to *F* and *F* to *M*	5 + 2 = 7
A to *F* and *F* to *S*	5 + 5 = 10

The smallest distance, 7, is the distance from *A to M*, so we write 7 beside *M*.

We now consider the towns *H*, *M* and *F*, and any roads leading from them to determine the shortest possible route.

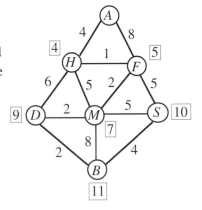

Route	Distance
A to *H* and *H* to *D*	4 + 6 = 10
A to *M* and *M* to *D*	7 + 2 = 9
A to *M* and *M* to *B*	7 + 8 = 15
A to *M* and *M* to *S*	7 + 5 = 12
A to *F* and *F* to *S*	5 + 5 = 10

The smallest distance, 9, is the distance from *A* to *D*, by going through *H* and *M*, so we write 9 beside *D*.

Following the same procedure, we write 10 beside *S*, and then 11 beside *B*.

The shortest distance from Amy's Cove to Burley is 11 km. The path to take is *AHFMDB* as illustrated in the diagram.

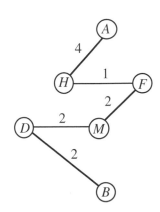

Suppose a mail carrier who lives at Amy's Cove has to visit the other six villages before returning home. How long is her shortest route?

From the route described above, we know that the only village she has missed is S. From B she can travel to S, then to F, then to H and finally to A for a total of 14 km.

Therefore the return path is $BSFHA$ and the total distance is $11 + 14 = 25$ km.

Example 2

When you reverse the digits of the number 24, the number increases by 18. How many two-digit numbers increase by 18 when their digits are reversed? Consider only numbers in which the number formed by reversing the digits is a two-digit number.

Solution

To increase a number by 18 means that the two-digit numbers differ by 18. To get a difference of 18 implies that the ones digits of the two numbers must differ by 8.

This is possible only if the two-digit numbers have the ones digits 0 and 2, 1 and 3, 2 and 4, 3 and 5, 4 and 6, 5 and 7, 6 and 8, 7 and 9, 8 and 0, or 9 and 1. For example, if the ones digits are 2 and 4, the numbers are 42 and 24. The possible numbers are:

$20 - 02 = 18$ (This is not possible, since 02 is not a two-digit number.)

$31 - 13 = 18$

$42 - 24 = 18$

$53 - 35 = 18$

$63 - 46 = 18$

$75 - 57 = 18$

$86 - 68 = 18$

$97 - 79 = 18$

$80 - 08 = 72$ (This is not possible, since 08 is not a two-digit number.)

$91 - 19 = 72$.

There are seven two-digit numbers that increase by 18 when their digits are reversed.

An alternate solution using algebra.

Note: A two-digit number such as 42 means $4(10) + 2$.

Let the tens digit of a two-digit number be represented by a and the ones digit by b.

The number is $10a + b$.

The number with the digits reversed is $10b + a$.

To increase a number by 18 means that the two-digit numbers differ by 18.

Therefore $(10b + a) - (10a + b) = 18$

$$10b + a - 10a - b = 18$$

$$9b - 9a = 18.$$

Divide by 9 to get $b - a = 2.$

That is, the ones digit is 2 greater than the tens digit.

There are seven possible two-digit numbers with this property: 13, 24, 35, 46, 57, 68, and 79.

Extending a problem

In chapter one we noted that some problems lead to extensions. For example, problem 9 from chapter one asked you to place the integers 1 to 8 around a square, with one at each vertex and one on each side, such that the sum of the numbers along each side is the same?

One solution is given in the answer section.   A natural question to ask is "How many other solutions are there?"

As part of the solution, we noted that the sum of the integers at the four corners is a multiple of 4 and that the possible sums of these four integers are 12, 16, 20, and 24.

How did we arrive at this conclusion?

Here is an algebraic explanation.

Let's place the variables *a, b, c,* and *d* at the four corners and the variables *e, f, g,* and *h* along the four sides of the square. These variables represent the integers from 1 to 8, in some order.

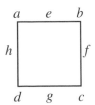

The sum of the integers from 1 to 8 is 36.   (Recall the formula from chapter 5.)
But from our diagram the sum is $a+b+c+d+e+f+g+h$.
This means that $a+b+c+d+e+f+g+h = 36$.

We also know that the sum along each side is the same.
Let this sum be k.

$$\text{Then} \qquad a+e+b=k$$
$$a+h+d=k$$
$$d+g+c=k$$
$$b+f+c=k.$$

Since each of these sums are the same, the four sums must be $4k$.
Let's add to get the equation
$$a+e+b+a+h+d+d+g+c+b+f+c=4k$$
or $a+b+c+d+e+f+g+h+a+b+c+d=4k$.

But $a+b+c+d+e+f+g+h = 36$, so we can write the equation as
$$36+a+b+c+d=4k.$$
This means that the sum of the four corners is
$$a+b+c+d = 4k-36.$$
Since $4k$ is a multiple of 4 and 36 is also a multiple of 4, then $4k-36$ is a
multiple of 4.

The variables a, b, c, and d can be replaced by any of the integers from 1 to 8.
Suppose we choose the smallest possible values for these variables, that is, 1, 2,
3, and 4.   Then $1+2+3+4=10$, so the smallest value that $a+b+c+d$ can
have is 10.
Suppose we choose the largest possible values for a, b, c, and d, that is, 5, 6, 7,
and 8.
Then $5+6+7+8=26$, so the largest value that $a+b+c+d$ can have is 26.
But $a+b+c+d = 4k-36$, so the value for $4k-36$ must be between 10 and
26. The only multiples of 4 between 10 and 26 are 12, 16, 20, and 24.

Let the sum of the four corners be 24.
$$\text{Then} \quad 4k-36=24$$
$$4k=60$$
$$k=15.$$

Therefore the sum of the three integers along each side is 15 and the sum of the four corners is 24.

How can we get a sum of 24 from the integers 1 to 8?

To get 24, we should begin with the largest integer and consider various combinations.

Note that $8 + 7 + 6 + 3 = 24$ and $8 + 7 + 5 + 4 = 24$.

Will any combination work if we begin with 7?

Well, $7 + 6 + 5 + 6 = 24$, but this is not possible since we have used 6 twice.

It appears that the only possible combinations are the two we have above.

Do both of these give a side sum of 15?

Case 1. Place the set 8, 7, 6, and 3 at the four corners.

Since the sum of the three integers along each side must be 15, the integers 8 and 7 cannot be placed as shown in the diagram.

Let's place the 8 and 7 at opposite corners. Eureka! This works. The sum of the integers on each side is 15.

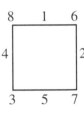

Case 2. Place the integers 8, 7, 5, and 4 at the four corners.

We know, from case 1, that we cannot place the integers 8 and 7 along the same row, so let's consider the situation in the diagram.

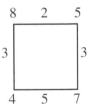

This means that we need to use the integers 3 and 5 twice and this is not possible.

Switching the integers 5 and 4 doesn't help.

Therefore there is only one possible arrangement with the sum along each side equal to 15.

So far we have two answers to the problem; the answer given in the solutions and the answer we found in Case 1 above. Let's return to the solution that was given in chapter one. In this solution we were given that the sum of the four corners was 20.

Then $4k - 36 = 20$

$$4k = 56$$

$$k = 14.$$

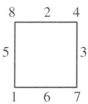

This means that the sum of the integers along each side is 14.

This represents the answer in the solution for problem 9 in chapter one.

Exercises

The following questions refer to problem 9 from chapter one.

1. Find another solution with $k = 14$ and the sum of the four corners equal to 20.

2. What is the sum of the three integers along each side of the square when the sum of the integers at the four corners is 12?

3. Find all possible solutions when the sum of the integers at the four corners is 12.

4. What is the sum of the three integers along each side of the square when the sum of the integers at the four corners is 16?

5. Find all possible solutions when the sum of the integers at the four corners is 16.

6. How many solutions are there to problem 9 in chapter one?

We talked earlier about the difference between exercises and problems. Here we have tried to show you that there are different types of mathematics problems. Some require that you use ideas developed in class in innovative ways. Others, as you have seen here, require that you develop completely new ways of thinking. These different approaches are what make mathematical problem solving challenging. Of course, they also are what make problem solving rewarding, because we sometimes see how to do a problem, when our classmates, and perhaps even our teachers, cannot. And that gives us a wonderful sense of achievement.

The problems in this set are a mixture of problems of all types, some you have met before, and some that are completely different to anything you have done. Don't expect to solve them all quickly. Take your time. Try different things. Above all, enjoy yourself.

After you have completed a problem, you should check the solutions. We have included numerous hints, suggestions and comments that we hope you will find interesting. From them you might discover an alternate method or a more efficient solution.

Problems

1.   What is the greatest number of Mondays that can occur in the first 45 days of a year?

2.   November has 30 days. November 12 falls on a Wednesday. Which day will occur five times during November?

3.   The mass limit for an elevator is 1500 kilograms. The average mass of the people in the elevator is 80 kilograms. If the combined mass of the people is 100 kilograms over the limit, how many people are in the elevator?

4.   When a certain number is divided by 9, the quotient is 6 and the remainder is 4. What is the number?

5. During the weekend, five people from Camp Gauss made phone calls. The cost and duration of their calls are indicated on the graph. Whose call cost the most per minute?

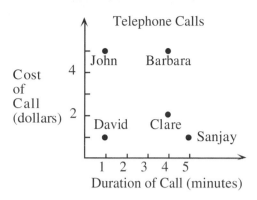

6. Daniel is constructing a ten-level tower. He puts ten blocks on the bottom level. In each level after that there is one block less than on the level below it. How many blocks are there in his completed tower?

7. Each kernel of magic popcorn always pops to become a perfect cube that is five times the volume of the original kernel. If each unpopped kernel has a volume of 1.6 cm^3, then what is the length of each edge of the pieces of the popped corn?

8. When Rami wrote the Gauss contest, he averaged 1 minute per question on the ten Part A questions and 2 minutes per question on the ten Part B questions. The contest has a time limit of one hour. What is the average amount of time he can spend on each of the five Part C questions?

9. A well is 20 m deep. A snail at the bottom climbs up 4 m each day and slips back 2 m each night. How long will it take for the snail to reach the top of the well?

10. What is the area of the whole figure, given that the shaded area is a square?
    (All the angles in the diagram are right angles.)

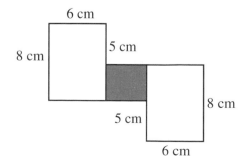

11. A ball is dropped from 60 cm and always bounces $\frac{2}{3}$ of the height from which it falls. How many centimetres will the ball rise after the third bounce?

12. A carton of whipping cream contains 35% butterfat. A carton labelled "2% milk" contains 2% butterfat. If a volume of whipping cream is mixed with an equal volume of 2% milk, what is the percentage of butterfat in the mixture?

13. In a game at a local fair, you draw one yellow card and one blue card. Each yellow card has one of the numbers 1, 3, 5, 7, 9, or 12 printed on it. Each blue card has one of the numbers 1, 2, 3, 4, 5, or 6 printed on it. If the product of your two numbers is a perfect square, you have drawn a winning pair. How many winning pairs are there?

14. Ten points are spaced equally around a circle. How many different chords can be formed by joining any two of these points? (A chord is a straight line joining two points on the circumference of a circle.)

15. In a ring toss game at a carnival, three rings are tossed over any of three pegs. A ring over peg *A* is worth *one* point, over peg *B three* points, and over peg *C five* points. If all three rings land on pegs, how many different point totals are possible? (It is possible to have more than one ring on a peg.)

16 A field trip is being planned for some Grade 8 students. The number of people going on the trip, including students, teachers and parents, is 1161. Each bus can hold at most 49 passengers. If all buses have the same number of passengers, what is the least number of buses needed for the field trip?

17. A parade marches through town from the Mall (M) to the Community Centre (CC). If the parade can only travel east or south, how many different possible routes are there?

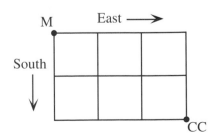

18. Ten teenagers living in ten different houses decide to install their own private telephone system. The system is to provide a direct link between each pair of the ten houses. A separate line is required for each link. What is the minimum number of lines required?

19. A watch keeps exact time, but it has only an hour hand. When the hour hand is $\frac{2}{5}$ of the distance between the "4" and the "5", what is the correct time?

20. When a pitcher is $\frac{1}{2}$ full it contains exactly enough water to fill three identical glasses. How full would the pitcher be if it had exactly enough water to fill four of the same glasses?

21. In a skate-a-thon, for every lap Frances skates, Sangeeta skates two laps. For every lap Sangeeta skates, Timoula skates four laps. If Frances skates five laps, then what is the total number of laps skated by these three skaters?

22. A ride on the "Quadzilla chair" at the Panorama Ski Resort takes you to a 750 m higher elevation in 8 minutes. If the horizontal distance from the bottom of the mountain to the start of the lift is 1000 m, what is the speed, in metres per minute, of the chair along its supporting cable?
Write your answer correct to one decimal place.

23. Of the 250 students at a basketball game, 185 are wearing blue jeans, 110 have blue eyes, and 55 are neither wearing blue jeans nor have blue eyes. How many students who are wearing blue jeans have blue eyes?

24. On a 800 km trip, a family travelled at 90 km/h for the first 600 km and 100 km/h for the remainder of the journey. If they had travelled 90 km/h for the entire journey, how much longer would the trip have taken?

25. Rearranging the digits 1, 2, 3, and 4, it is possible to form twenty-four different four-digit numbers, including the number 1234. What is the sum of these twenty-four numbers?

26. The diagram is a road map of roads linking the two villages of Amy's Cove to Burley, passing through Hillsdale, Fractionville, Deepwell, Manhill, Steamroller, Greenlee, and Rowntree.
The number on each section of road is the length in kilometres of that road.
How long is the shortest route from Amy's Cove to Burley?

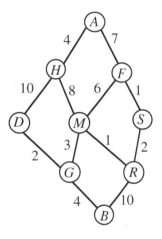

Suppose a mail carrier who lives at Amy's Cove has to visit the other eight villages before returning home. How long is his shortest route?

27. The holes on the peg board shown are one unit apart, both horizontally and vertically. Pegs *X* and *Y* are positioned as shown In how many holes can a peg *Z* be inserted so that *XYZ* forms a right-angled triangle?

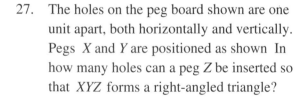

28. A spider and a caterpillar travel clockwise around the edge of a rectangular table whose length *RQ* is twice its width *PQ*. The spider starts at *P* when the caterpillar starts at *R*. The spider overtakes the caterpillar when the caterpillar reaches *Q* for the first time. If the caterpillar crawls at a speed of 2 mm/s, what is the speed at which the spider crawls?

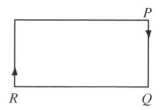

29. Each of the numbers from 1 to 9 is placed, one per circle, into the pattern shown. The sums along each of the four sides are equal. How many entries are possible for the middle circle?

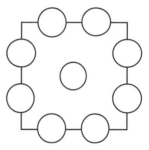

30. The sum of nine consecutive positive integers is 99. What is the largest of these integers?

31. Each of the 12 edges of a cube is coloured either red or green. Every face of the cube has at least one red edge. What is the smallest number of red edges?

32. Rearrange three of the toothpicks to form three squares of the same size.

33. *A, B* and *C* are houses and *E, G* and *W* represent sources of electricity, gas and water. Can you connect each house with each utility so that the lines do not cross each other.

34.   In a grade 8 class of 380 students, 70% can swim, 60% of the boys can swim, and 200 girls can swim. How many girls are in the class?

35.   (a)   How many 2 cm by 2 cm by 2 cm cubes will fit into a 4 cm by 4 cm by 4 cm box?

     (b)   How many of the 2 cm cubes will fit into a 7 cm by 7 cm by 7cm box?

36.   Twenty-seven identical white cubes are stacked together to form one larger cube. All the outside faces of the larger cube are then painted red. If the larger cube is taken apart, how many of the smaller cubes have exactly four white faces?

37.   Two standard dice, one yellow and one blue, are rolled together four times. The number showing on the yellow die increases on each roll, while the number showing on the blue die decreases on each roll. What is the smallest possible sum on the second roll?

38.   (a)   The letters of the word 'GAUSS' are cycled (meaning that the first letter is moved to the back), producing AUSSG, USSGA, and so on, with each new version written on the next line. In the same way, on the same line, the digits of the number 1234 are cycled. The first two lines are

              1. GAUSS   1234
              2. AUSSG   2341.

    On what line will GAUSS 1234 first appear again?

     (b)   Using the word GEOMETRY instead of GAUSS and the number 123456, on what line will GEOMETRY 123456 appear for the third time (counting the original line)?

39.   If the sides of a triangle have lengths 30, 40 and 50, what is the length of the shortest altitude?

40. Stephen had a 10:00 a.m. appointment 60 km from his home. He averaged 80 km/h for the trip and arrived 20 minutes late for the appointment. At what time did he leave home?

41. A diagonal cross is drawn on a 10×10 square as shown in the diagram. What is the area of the cross?

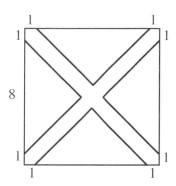

42. Paving stones come in the three sizes indicated in the diagram. Equal numbers of each size are used to make a single square pattern. What is the total number of stones required to form the smallest possible square?

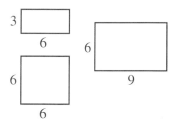

43. Without lifting your pencil from the page, draw four straight lines that pass through all nine points.

44. (a) Using the integers from 1 to 9, place one integer in each of the circles, so that the sum along each side of the triangle is 20. Use each number only once.

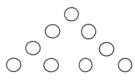

(b) Use the same numbers to obtain a sum of 17 along each side.

(c) What other sums are possible along each side?

45. A farmer is on the way to market. He has with him a goose, a fox and some wheat. He comes to a river that may only be crossed by means of a rickety old rowboat that is so unsafe he can only take one item at a time in the boat with him. (In addition he must ensure that the fox doesn't eat the goose, or the goose doesn't eat the wheat.) If the river is 50 m wide, what is the least distance the farmer must row in order to get the goose, the fox and the wheat across the river?

46. A cubic block of cheese is to be cut into eight identical pieces. What is the least number of knife cuts required?

    In how many ways can this be done?

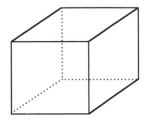

Answers and Solutions

Chapter 1   *Introduction to Problem Solving*
Problems

1. The string could be divided into four equal parts. Cut the string one quarter of the way from one end. Then one piece will be $\frac{3}{4} \times 72 = 54$ cm long and the other piece will be 18 cm long.

2. The number of posts required is $236 \div 4 + 1 = 60$.

3. All of the numbers are multiples of 10, so any combination of them, when added, will give a multiple of 10.
    Since 75 is not a multiple of 10, it is impossible for any combination of the given numbers to add to 75.

4. In one minute there are $60 \div 5 = 12$ five-second intervals.
    Therefore $3 \times 12 = 36$ drops will fall in one minute.

5. We can make a chart to show the time, in seconds, that each elevator arrives at and departs from each floor.

Floor	Fast elevator		Slow elevator	
	Arrival Time	Departure Time	Arrival Time	Departure Time
6		0		0
5	10	$10 + 30 = 40$	20	20
4	50	50	40	$40 + 30 = 70$
3	60	$60 + 30 = 90$	90	90
2	100	$100 + 30 = 130$	110	110
1	140		130	

The slow elevator reaches the ground floor first.

6. The sum of the vertical distances is $2 \times 20 = 40$ cm.
    The sum of the horizontal distance is 18 cm.
    Therefore the length of the solid line is $40 + 18 = 58$ cm.

7. If we add the dotted lines to complete
    the rectangle, as shown, the distance
    around the figure is the same as the
    perimeter of the rectangle.
    Since the dimensions of the rectangle
    are 16 cm by 13 cm, the distance around
    the figure is $2(16+13) = 58\,\text{cm}$.

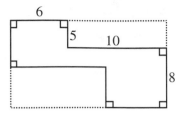

8. The key to solving this problem
    is to note that the circle labelleled
    A has the most lines connecting
    to other circles. Since 1 and 6
    each have only one adjacent
    number, we should place one of
    them, say 6, in A.

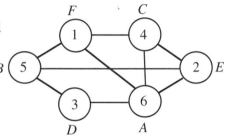

Then 5 must go in B, 4 in C, 3 in D, 2 in E or F, and 1 in the
remaining circle.
Try to convince yourself that it is impossible to start with 2, 3, 4, or 5 in
circle A.

9. One of several solutions is shown in the
    diagram.
    It is interesting to note that the sum of the
    integers at the four corners is 20 which is a
    multiple of 4. In fact, it can be shown that
    the sum of the corner integers must be a
    multiple of 4 and the only possibilities are
    12, 16, 20, and 24.
    With this much information you should
    see how many solutions you can find.

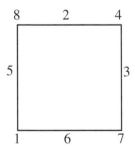

10. Statement (iv) tells us that Sandy is shorter than Ron who, in turn, is shorter than Alicia.

Statement (ii) tells us, in addition, that Alicia is shorter than Ed.

Statement (v) tells us that Ed is shorter than Gurinder who, in turn, is shorter than Jane.

Therefore the ranking of the students, from shortest to tallest, is Sandy, Ron, Alicia, Ed, Gurinder, and Jane.

Notice that we don't need statements (i) and (ii) to do this ranking.

11. The longest path is *ACDECB*, a total of 5 units.

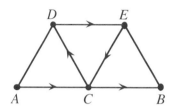

12. Suppose we begin by colouring the middle circle red. Then we can colour alternate circles around the outside green and blue, as illustrated.

Thus we need a minimum of three colours.

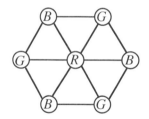

13. The sum of all the numbers in the circle is $1+2+3+...+11+12 = 78$. Then the sum of the number in each area is $78 \div 3 = 26$.

The diagram illustrates the positions of the two lines.

14.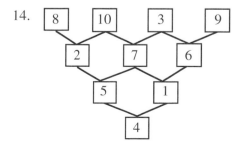

Chapter 2   *Numeracy*
Exercises

2.   2, 3, 5, 7, 11, 13, 17, 19, 23, 29, 31, 37, 41, 43, 47, 53, 59, 61, 67, 71, 73, 79, 83, 89, 97

3.   (a) 593 is prime.   (b) $921 = 3 \times 307$, so 921 is not prime.
      (c) 2789 is prime.

6.   (a) $3 \times 3 \times 7$ (b) $2 \times 5 \times 7$ (c) 3×41 (b) $3 \times 3 \times 5 \times 5 \times 7$

7. $5757 = 3 \times 19 \times 101$. The two greatest prime factors of 5757 are 19 and 101.

8. $(6+4) \times 5 \times (3-2+1)$ and $(5+6-1) \times (4 \times 3 - 2)$ are two possibilities.

9. $(5 \times 4) \times [(6 \times 2) \div 3 + 1]$ and $(6 \times 3 + 4 - 2) \times 5 \times 1$ are two possibilities.

10. $(6 \times 4 + 1) \times (5 + 2 - 3)$ and $(6 \times 3 + 5 + 2) \times 4 \times 1$ are two possibilities.

11. $(6+4)^2 (5-3-1)$ and $(6+4)^3 \div (5 \times 2 \times 1)$ are two possibilities.

12. $(9+4-3)^{6 \div 2}$ and $\left(9 + 3 - \sqrt{4}\right)^{6 \div 2}$ are two possibilities.

13. 6   14. 6   15. (a) not necessarily   (b) yes

16.   (a)   Divisors are 1, 2, 3, 4, 6, 8, 12, 24.
          Factors are 1×24, 2×12, 3×8, 4×6, $2 \times 2 \times 6$, $2 \times 3 \times 4$, and $2 \times 2 \times 2 \times 3$.

      (b)   Divisors are 1, 3, 9, 27.
          Factors are 1×27, 3×9 and $3 \times 3 \times 3$.

      (c)   Divisors are 1, 2, 4, 13, 26, 52.
          Factors are 1×52, 2×26, 4×13, and $2 \times 2 \times 13$.

17. (a) 14   (b) 15   (c) 42   18. $2 \times 2 \times 2 \times 3 \times 5 = 120$

19. (a) 84   (b) 60   (c) 1260   20. $\frac{19}{40}$

21. (a)   The length of each side of the smallest square is the LCM of 24 and 39, namely, 312.

     (b) $8 \times 13 = 104$ rectangles will be required.

Solutions to Problems

1.   We need not consider any primes since a prime has only two positive divisors.

     The divisors of 14 are 1, 2, 7, and 14; the divisors of 15 are 1, 3, 5, and 15; and the divisors of 16 are 1, 2, 4, 8, and 16. None of these has six divisors but 18 has six divisors, namely, 1, 2, 3, 6, 9, and 18. The factors of 18 are 1×18, 2×9, 3×6, and $2 \times 3 \times 3$.

2.   Factoring 315 into primes, we find that $315 = 3 \times 3 \times 5 \times 7$.

     The factorizations of 315 as the product of two integers are 3×105, 5×63, 7×45, 9×35, and 15×21. There are five possible factorizations.

3.   Factoring 7315 into primes, we get $7315 = 5 \times 7 \times 11 \times 19$.

     Then 7315 can be written as the product of the two-digit integers 95 and 77.

4.   (a)   *Solution 1*

            The two-digit house numbers that can be formed are 35, 37, 53, 57, 73, and 75.

            Thus six two-digit house numbers can be formed.

            Solution 2

            There are three choices for the tens digit of the house number, namely, 3, 5 or 7.

            For each of these choices, there are two choices remaining for the units digit.

            Thus six two-digit house numbers can be formed.

     (b)   Since the numbers are all odd, none of them is divisible by 6.

5. $1 + 2 - 3 - 4 + 5 + 6 - 7 - 8 + \cdots + 45 + 46 - 47 - 48 + 49$
     $= 1 + (2 - 3 - 4 + 5) + (6 - 7 - 8 + 9) + \cdots + (46 - 47 - 48 + 49)$
     $= 1 + 0 + 0 + \cdots + 0 + 0$
     $= 1.$

Note: There are many ways of reorganizing the numbers to find the sum.

6.   *Solution 1*
     After the 10% increase on Monday, the value of the stock was 110% of
     $\$100 = 1.1 \times \$100 = \$110$.
     After the 10% decrease on Tuesday the value of the stock was 90% of
     $\$110 = 0.9 \times \$110 = \$99$.
     After the 10% increase on Wednesday the value of the stock was 110% of
     $\$99 = 1.1 \times \$99 = \$108.90$.
     This represents a gain of \$8.90 over the three-day period.
     This is an increase of $\frac{8.9}{100} \times 100 = 8.9\%$.

     Solution 2
     The value of the stock at the end of the three-day period is
     $1.1 \big[0.9 (1.1 \times \$100) \big] = (1.1)(0.9)(1.1)(\$100) = \$108.90$.
     The percentage increase over this period is $\frac{8.9}{100} \times 100 = 8.9\%$.

7.   Since $5913d8$ is divisible by 12 it must be divisible by 3 and by 4. To be
     divisible by 3, the sum of its digits must be divisible by 3.
     Therefore $5 + 9 + 1 + 3 + d + 8 = 26 + d$ must be divisible by 3; so d must
     be 1, 4 or 7.
     For the number to be divisible by 4, the number formed by the last two
     digits must be divisible by 4.
     By checking 18, 48 and 78, we find that only 48 is divisible by 4.
     Therefore $d = 4$.

8.   Three answers can be obtained, namely, $n = 1 + 2 - 3 = 0$, $n = 1 + 3 - 2 = 2$
     and $n = 2 + 3 - 1 = 4$.
     If the numbers 1, 2, 3, and 4 are used the possible values of n are -1, 0, 1,
     2, 3, 4, 5, and 6, a total of 8.

9.   I gave $\frac{3}{4} \times \frac{1}{3} = \frac{1}{4}$ of the pizza to my friend.

This left $1 - \frac{1}{3} - \frac{1}{3} - \frac{1}{4} = \frac{1}{12}$ of the pizza for me.

10. If we pair up terms from the two brackets we get

$$(4 + 8 + 12 + 16 + \cdots + 256) - (1 + 5 + 9 + 13 + \cdots + 253)$$
$$= (4 - 1) + (8 - 5) + (12 - 9) + (16 - 13) + \cdots + (256 - 253)$$
$$= 3 + 3 + 3 + 3 + \cdots + 3$$
$$= 64 \times 3 \qquad \text{(Why are there 64 threes?)}$$
$$= 192.$$

11. When the tank is $\frac{1}{3}$ full it contains $\frac{1}{3} \times 48 = 16$ litres.

When the tank is $\frac{3}{4}$ full it contains $\frac{3}{4} \times 48 = 36$ litres.

Therefore, when the tank is $\frac{1}{3}$ full, Kathy must add 20 litres to make it $\frac{3}{4}$ full.

12. Since the numbers are all odd the product must be odd, and since one factor is 5, the units digit in the product is a 5.

13. The numbers of the pages showing are consecutive integers. The page numbers are 116 and 117 since $116 + 117 = 233$. The product of the page numbers is $116 \times 117 = 13\,572$.

14. The least score Stu could have had was $3 \times 3 = 9$ and the greatest score he could have had was $3 \times 7 = 21$. By replacing each score by an adjacent score, one at a time, there is a gain or loss of two points. Thus it is possible to achieve all odd integer scores between 9 and 21. The number of scores Stu could have had is 7.

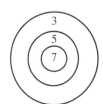

15. Since $67\,184 = 2 \times 2 \times 2 \times 2 \times 13 \times 17 \times 19$, and Don's children are all teenagers, their ages are 13, 16, 17, and 19.

16. *Solution 1*

    We are given that $6 + 7 + 8 + 9 + \cdots + 482 + 483 = 116\,871$.

    Hence, $8 + 9 + 10 + 11 + \cdots + 484 + 485 = 116\,871 - 6 - 7 + 484 + 485$
    $$= 117\,827.$$

    Solution 2

    In the second sum, each number is two larger than the corresponding number in the original sum.

    The number of numbers being added, in each case, is $483 - 5 = 478$.

    Thus $8 + 9 + 10 + 11 + \cdots + 484 + 485 = 116\,871 + 478 \times 2$
    $$= 116\,871 + 956$$
    $$= 117\,827.$$

17. The sums of the three rows are 46, 39, and 53, respectively.

    Since the average of the three rows is

    $\dfrac{46 + 39 + 53}{3} = 46$, the first row should not be

    changed. It is now necessary to increase the row whose sum is 39 by 7 and to decrease the row whose sum is 53 by 7. This can be done by interchanging 11 and 4.

16	8	22
23	4	12
11	27	15

18. *Solution 1*

    Since the sum of the six numbers in the row and column is 26, and because the five given numbers add to 20, then 6 must be added to bring the total to 26.

    Thus 6 must occur twice and is the number in the middle square.

    Solution 2

    The only possible combinations of numbers which add to 13 are $2 + 5 + 6$ and $3 + 4 + 6$.

    Therefore 6 must occur twice.

19.   If Chris divided by 8 and got 4, his previous number must have been 32.
He added 2 to get 32 so he must have started with 30.
If he had taken 30, added 8 and then divided by 2, he would have
obtained $(30+8) \div 2 = 38 \div 2 = 19$.
Therefore the final answer would have been 19.

20.   The smallest difference will be produced when the three-digit number is
as small as possible, that is, 356, and the two-digit number is as large as
possible, that is, 87.
The smallest difference is $356 - 87 = 269$.

21.   Since the sum of the five integers is 105, their average is $\frac{105}{5} = 21$.
Because there is an odd number of consecutive integers, 21 is the middle
integer and 23 is the largest.

22.   Since there are three bundles, each with 100 bills in them, the three
bundles would be worth $500, $1000 and $2000, respectively.
Since there are 10 bundles of each type of bill, their overall value would
be $10(\$500 + \$1000 + \$2000) = \$35\,000$.

23.   When the figure is folded to make a cube,
the numbers 8 and 6 are on opposite faces
so that it is *not* possible to achieve $8 \times 7 \times 6$
or 336. It is possible, however, to have the
sides with 5, 7 and 8 to meet at a corner
which gives the answer $5 \times 7 \times 8$ or 280.

24.   If 100 g of white rice is added to the mixture, then the new mixture
contains 140 g of white rice and 60 g of brown rice.
The percentage of the new mixture that is white rice is $\frac{140}{200} = 0.70$
or 70%.

Chapter 3   *Geometry*
Exercises

1.   0, 1, 3, 4, 5, 6

2.   Alternate angles: c and f, d and e.
      Corresponding angles: a and e, c and g, b and f, d and h.

3.   (a) $x = 26$ (b) $y = 25$ (c) $y = 110$, $x = 35$ (d) $x = 61$

      (e) $x = 30$, $y = 21$ (f) $y = 110$, $x = 40$

4.   (a) $x = 95$ (b) $x = 50$, $y = 50$ (c) $x = 25$, $y = 80$

5.   In $\triangle ACE$, $\angle A + \angle C + \angle E = 180°$. Similarly, in $\triangle BDF$,
      $\angle B + \angle D + \angle F = 180°$.
      It follows that $\angle A + \angle B + \angle C + \angle D + \angle E + \angle F = 360°$.

6. $\angle PRS = 60°$ and $\angle PSR = 70°$.
      Therefore $\angle RPS = 180° - 60° - 70° = 50°$.

7.   (a) 540°   (b) 720°   (c) 900°   (d) 1080°   (e) 1800°

8.   For a 50-sided polygon, the sum of the interior angles is
      $50 \times 180° - 2 \times 180° = 48 \times 180° = 8640°$.

      If a point P is selected inside an n-sided
      polygon and this point is joined to all the
      vertices we obtain n triangles. The sum of
      the angles in these n triangles is $180n$
      degrees. But the sum of the angles around
      P is 360°. Therefore the sum of the
      interior angles in an n-sided polygon is
      $180n - 360$ or $180(n - 2)$ degrees.

9.   In $\triangle ABC$, $AB + BC > AC$ (the triangle inequality). Since the rope has
      been pulled 1 m, then $BC = 11$ m. Therefore $AB + 11 > 12$; so $AB > 1$,
      and the canoe moves more than 1 m.

10.　(a) 25　　　　(b) $y^2 + 64 = 400$; $y = \sqrt{336} \doteq 18.3$

11.　Using the Pythagorean Theorem, we get

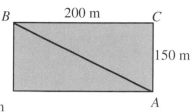

$$AB^2 = 150^2 + 200^2 = 62\,500$$

$$AB = \sqrt{62\,500} = 250.$$

One must walk $200 + 150 - 250 = 100$ m further.

12.　Let the length of each side of the square beam be x cm.

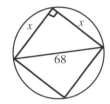

Then, $x^2 + x^2 = 68^2$

$$2x^2 = 4624$$

$$x^2 = 2312$$

$$x = \sqrt{2312}$$

$$\doteq 48.1.$$

The largest square beam that can be cut from the log has side length 48 cm.

Solutions to Problems

1.　In each 5-minute interval the minute hand rotates through an angle of $360° \div 12 = 30°$.
Between 9:15 a.m. and 9:35 a.m., the minute hand rotates $4 \times 30° = 120°$.

2. In isosceles triangle ABC,

    $\angle ABC = \angle ACB = \frac{1}{2}(180° - 40°) = 70°$.

    In $\triangle ABD$, $\angle ABD = 180° - 90° - 40° = 50°$.
    Therefore $\angle DBC = 70° - 50° = 20°$.

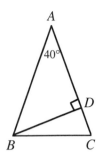

3. In the diagram, two additional angles
    are assigned measures a and b.
    Since $130 + b = 180$, $b = 50$.
    Since the sum of the three interior
    angles of a triangle is $180°$,
    $a = 180 - 50 - 60 = 70$.
    Finally, $x + 70 + 80 = 180$.
    Thus $x = 30$.

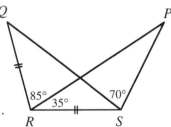

4. Since $RQ = RS$, triangle QRS is
    isosceles.
    In triangle QRS,
    $\angle QRS = 85° + 35° = 120°$.

    Thus $\angle RQS = \angle RSQ = \dfrac{180° - 120°}{2} = 30°$.

    In triangle PRS,
    $\angle RPS = 180° - (35° + 30° + 70°)$
    $= 180° - 135°$
    $= 45°$.

5. The number of degrees through which
    the hour hand rotates in one hour is
    $\frac{360°}{12} = 30°$.
    The elapsed time from 9:20 a.m. to
    11:50 a.m. is $2\frac{1}{2}$ hours.

The angle through which the hour hand travels in this time interval is

$2\frac{1}{2} \times 30° = 75°$.

6. The fifteen spokes produce fifteen equal angles.

    Therefore the size of each angle is $\frac{360°}{15} = 24°$.

7. At 11:00 a.m., the angle between the hands is $\frac{1}{12} \times 360° = 30°$. Between 11:00 a.m. and 11:18 a.m., the minute hand rotates $\frac{18}{60} \times 360° = 108°$.

    In each hour, the hour hand rotates

    $\frac{1}{12} \times 360° = 30°$.

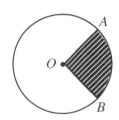

    Between 11:00 a.m. and 11:18 a.m., the hour hand rotates $\frac{18}{60} \times 30° = 9°$.

    The angle between the hands at 11:18 a.m. is $30° + 108° - 9° = 129°$.

8. Since the shaded area represents 20% or $\frac{1}{5}$ of the area of the circle, then

    $\angle AOB = \frac{1}{5}(360°) = 72°$.

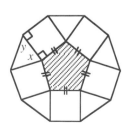

9. *Solution 1*
    The sum of the interior angles of a pentagon is $180°(5-2) = 540°$.
    Since the angles in a regular pentagon are all equal, each one is $540° \div 5 = 108°$.
    The sum of the angles of each vertex of the pentagon is $x + 90° + 90° + 108° = 360°$.
    Therefore $x = 72$.

Solution 2

The outside figure has 10 sides, so the sum of all its interior angles is $8 \times 180° = 1440°$.

Since all 10 angles are equal, each one is $144°$.

Then $y + 90 = 144$

$$y = 54.$$

Since $x + 54 + 54 = 180$, $x = 72$.

10.   In $\triangle ABC$, $\angle ABC = \angle ACB$

$$= \frac{1}{2}(180° - 40°)$$

$$= 70°.$$

Since $2x + 70 = 180$

$$2x = 110$$

$$x = 55.$$

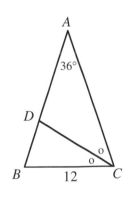

11.   In $\triangle ABC$, $\angle ABC + \angle ACB = 180° - 36° = 144°$.

Since $AB = AC$, $\angle ABC = \angle ACB = 72°$.

Therefore $\angle ACD = \frac{1}{2} \times 72° = 36°$.

In $\triangle CBD$, $\angle CDB = 180° - 72° - 36° = 72°$.

Therefore $\triangle CDB$ is isosceles and $CD = BC = 12$.

12.   In the diagram, sectors DOC, EOF and AOB are all equal.

Thus $\angle EOF = \angle AOB = y$.

Similarly, sectors DOE, FOA and BOC are equal, so $\angle DOE = \angle BOC = x$.

Since these sectors form a complete circle, $3x + 3y = 360$, or $x + y = 120$.

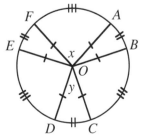

13. Since $b + 130 = 180$, $b = 50$.
     Since $x + 70 = 180$, $x = 110$.
     Therefore $a = 180 - 50 - 110 = 20$.

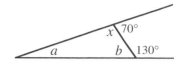

14. Since $QRST$ is a square, $\angle TQR = 90°$.
     Since $\triangle PQT$ is equilateral, $\angle PQT = 60°$.
     Thus $\angle PQR = 90° + 60° = 150°$.
     Since $PQ = QR$, triangle QPT is isosceles and

$$\angle QPR = \angle QRP$$
$$= \frac{1}{2}(180° - 150°)$$
$$= 15°.$$

     Therefore $\angle QRP = 15°$.

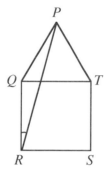

15. Extend PQ to R and label the angles in
     the diagram, as shown.
     Since $y + 150 = 180$, $y = 30$.
     Similarly, $z = 40$ and $w = 50$.
     In $\triangle PRS$, $\angle PRS = 180° - y - z = 110°$.
     Therefore $v = 70$.
     Then $x = 180 - v - w = 180 - 70 - 50 = 60$.

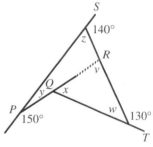

Note: The sum of the angles of quadrilateral $PSTQ$ is $360°$.
         Therefore reflex $\angle PQT = 360° - 30° - 40° - 50° = 240°$.
         Then $x = \angle PQT - 180 = 240 - 180 = 60$.

16. In isosceles triangles DBC and ADB, label
     the equal angles, as shown.
     In $\triangle ABD$, $2y + z = 180$.
     Since $x + z = 180$, then $x = 2y$.
     Then $5y = 180$
     $\qquad\quad y = 36$.
     Therefore $\angle ACB = 2(36°) = 72°$.

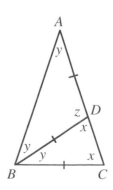

17. Let the measures of angles S and Q be x and $4x$ respectively.
Since the sum of the angles of quadrilateral $PQRS$ is $360°$, then

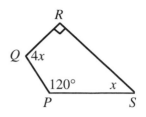

$$4x + 120 + x + 90 = 360$$
$$5x + 210 = 360$$
$$5x = 150$$
$$x = 30.$$

Therefore $\angle S = 30°$.

18. Since $(2w - 40) + 50 = 180$
$$2w = 170$$
$$w = 85.$$

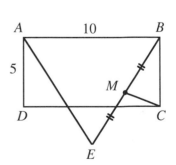

Since AEK and CGE are corresponding angles, $\angle CGF = 75°$.
Then $y = 105$.
Since EFH and GHJ are corresponding angles, $\angle GHJ = 50°$
and so $z = 130$.
Since $y = 105$, $\angle JGH = 75°$.
Therefore $x = 180 - 75 - 50$
$$= 55.$$

19. Since ABE is an equilateral triangle, $AB = BE = 10$.
Since M is the midpoint of BE, then $BM = 5$.
Since $BC = 5$, $\triangle BMC$ is isosceles and $\angle BMC = \angle BCM$.
But $\angle ABC = 90°$ and $\angle ABE = 60°$, so $\angle MBC = 30°$.
Then $\angle BMC = \angle BCM$

$$= \frac{1}{2}(180° - 30°)$$
$$= 75°.$$

Chapter 4   *Measurement*
Exercises

1. $13+13+6+6+3+3 = 44$ cm

2.   The third side cannot be 3 cm since it is not possible to form a triangle with sides 3 cm, 3 cm and 8 cm. (Why?). Therefore the perimeter is $8+8+3 = 19$ cm.

3.   Since half the perimeter is 48 cm and the width is 15 cm, the length of the rectangle is $48-15 = 33$ cm.

4.   Let the sides be x cm, $(8+x)$ cm and $(8+x)$ cm.

$$x+(8+x)+(8+x) = 49$$
$$3x+16 = 49$$
$$3x = 33$$
$$x = 11.$$

Each of the equal sides is 19 cm long.

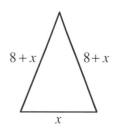

5.   The two semicircular ends of the track form a circle whose diameter is 30 and whose circumference is

$$30\pi \doteqdot 30 \times 3.14$$
$$= 94.2.$$

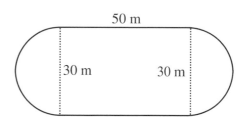

The perimeter of the track, to the nearest metre, is $94+50+50 = 194$ m.

6. $30 \times 50\pi \doteqdot 4710$ cm or 47.1 m.

7.   The circumference is also doubled.

8.   4 cm   9. $6 \times 10 - 6 \times 3 = 42$ cm^2.

10. Not usually.   11. All sides must be equal.

12. $\frac{1}{2}(11+15) \times 7 = 91$ cm^2.   13. 3 cm^2

14. Quadrilateral *AEFD* is a trapezoid with area $\frac{1}{2}(1+3)\times 4 = 8\text{ cm}^2$.
    Alternatively, note that *AEFD* and *CFEB* are identical trapezoids, each
    with half the area of the square. So the area of $AEFD = \frac{1}{2}\times 4^2 = 8\text{ cm}^2$.

15. $16\pi \doteq 50.2\text{ cm}^2$

16. Since the area is 36 cm^2, the side length is 6 cm. Therefore the
    perimeter is 24 cm.

17. The area of $\triangle ABC$ is $\frac{1}{2}\times 12\times 5 = 30\text{ cm}^2$.
    The area of $\triangle ABD$ is $30-20 = 10\text{ cm}^2$.

19. We can use Heron's Formula to find the areas of triangles *ADB* and
    DBC.
    Area of $\triangle ADB = \sqrt{6\times 1\times 1\times 4} = \sqrt{24} \doteq 4.9\text{ cm}^2$.
    Area of $DBC = \sqrt{8\times 3\times 3\times 2} = \sqrt{144} = 12\text{ cm}^2$.
    Area of $ABCD = 4.9+12 = 16.9\text{ cm}^2$.

20. 216 cm^2 21. 1080 cm^2 22. $16\pi\text{ cm}^2$ or 50.2 cm^2

23. The area of the base is $12\times 12 = 144\text{ cm}^2$.
    The area of each triangular face is $\frac{1}{2}\times 12\times 20 = 120\text{ cm}^2$.
    The surface area of the pyramid is $144+4\times 120 = 624\text{ cm}^2$.

24. (a) $6\pi\times 15 = 90\pi \doteq 283\text{ cm}^2$.
    (b) $2\times\pi\times 3^2 + 90\pi = 108\pi \doteq 339\text{ cm}^2$.

25. (a) The area of the circle is

$$\pi \times 12^2 = 144\pi \text{ cm}^2.$$

The area of the sector after the shaded portion is removed is

$$\frac{300}{360} \times 144\pi = 120\pi \doteq 377 \text{ cm}^2.$$

12 cm

(b) The circumference of the top of the cone is equal to the remaining part of the circumference of the circle after the shaded part is removed; that is, $\frac{300}{360} \times 24\pi = 20\pi \doteq 63 \text{ cm}$.

26. $10^3 - 5^3 = 875 \text{ cm}^3$.   27. 24   28. 25 cm^2

29. 174 cm^3 30. Yes   31. $25\pi \times 20 = 500\pi \doteq 1570 \text{ cm}^3$.

Solutions to Problems

1. Each square has area 100 cm^2. Then the side of each square has length 10 cm and so its perimeter is 40 cm.

2. Since $\triangle ADE$ is right-angled, $AD = \sqrt{4^2 + 3^2} = 5$.
   Similarly $BC = 5$. The perimeter of the trapezoid is $2(5 + 4 + 8) = 34$ cm.

3. Draw $QM \perp SR$. Then $QM = 12$ cm and $MR = 5$ cm.
   Using the Pythagorean Theorem, we get $QR = \sqrt{12^2 + 5^2} = 13$.
   Also, the circumference of semicircle PTS is $\frac{1}{2} \times \pi \times 12 = 6\pi \doteq 19$ cm.

   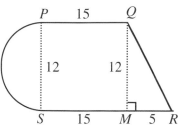

   The perimeter of the figure is $15 + 19 + 20 + 13 = 67$ cm.

4.   By the Pythagorean Theorem,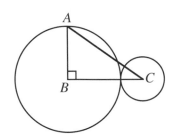

$$(AC)^2 = (AB)^2 + (BC)^2$$

$$100 = (AB)^2 + 64$$

$$36 = (AB)^2$$

$$AB = 6.$$

The radius of the larger circle is 6.
Thus the radius of the smaller circle is $BC - 6 = 8 - 6 = 2$ and its diameter is 4.
The circumference of the smaller circle is $\pi \times 4$ or 4π.

5.   The maximum number will occur when there is no waste cardboard, if that is possible. Since 25 is divisible by 5 and 28 is divisible by 4, there is no waste if cuts are made at 4 cm intervals along the 28 cm side and at 5 cm intervals along the 25 cm side. There will be $7 \times 5 = 35$ tickets.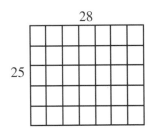

6.   The area of the shaded triangle is equal to the area of the square minus the area of the three unshaded triangles.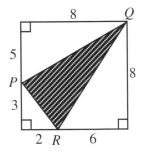

Hence the area of

$$\Delta PQR = 8 \times 8 - \left[\left(\tfrac{1}{2} \times 8 \times 5 \right) + \left(\tfrac{1}{2} \times 3 \times 2 \right) + \left(\tfrac{1}{2} \times 6 \times 8 \right) \right]$$

$$= 64 - [20 + 3 + 24]$$

$$= 64 - 47$$

$$= 17.$$

7. Divide the figure into a triangle and rectangle as shown.

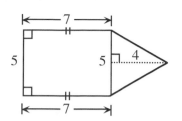

The required area is $(5)(7)+\frac{1}{2}(5)(4)=45$.

8. The total area of the four corner triangles is

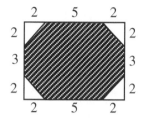

$4 \times \frac{1}{2} \times 2 \times 2 = 8$.

The shaded area is equal to the area of the 9×7 rectangle minus the area of the corner triangles; that is, $9 \times 7 - 8 = 55$.

9. The four curved parts of the band are each one-quarter of a circle with diameter 1 m. Their total length is $\pi \times 1$ m.

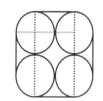

The length of each straight part of the band is the same as the distance between the centres of touching circles; that is, 1 m.
The length of the band is $\pi + 4 \doteq 7.14$ m.

10. Let the length of BC be x cm.
Thus the length of CD is $2x$ cm and the length of AB is $4x$ cm.

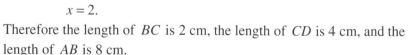

Since $AB + BC + CD = 14$ cm, we have

$$4x + x + 2x = 14$$
$$7x = 14$$
$$x = 2.$$

Therefore the length of BC is 2 cm, the length of CD is 4 cm, and the length of AB is 8 cm.

Since opposite sides are equal in length and $BC = DE$, the perimeter of the octagon is $4BC + 2CD + 2AB = 8 + 8 + 16 = 32$ cm.

11. In the diagram, we have illustrated a con-
venient way to draw the circle with its inner
and outer squares.

If we draw diameters PR and SQ, we have
the outer square divided into eight identical, or congruent, triangles.

Since the inner square has four of these
triangles, the area of the inner square is
one-half the area of the outer square.

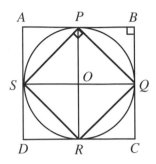

12. The area of the original rectangle is
$40 \times 30 = 1200 \text{ cm}^2$.
The length of the new rectangle is 120% of
40 cm or $1.2 \times 40 = 48$ cm.
The width of the new rectangle is 80% of
30 cm or $0.8 \times 30 = 24$ cm.
The area of the new rectangle is
$48 \times 24 = 1152 \text{ cm}^2$.
The difference in area between these two rectangles is
$1200 - 1152 = 48 \text{ cm}^2$.

13. The centre of the circle is joined to the
endpoints of the chord to produce an equilateral triangle of side length 12 cm.
Hence $\angle BOA = 60°$ and the arc AB has a
length which is $\frac{60}{360} = \frac{1}{6}$ that of the
circumference of the circle.
The circumference of the circle is $(2)(12 \text{ cm})(\pi) = 24\pi \text{ cm}$.
Thus the perimeter of the shaded segment, in cm, is
$12 + \frac{1}{6}(24\pi) = 12 + 4\pi \doteq 24.6$.

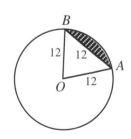

14. Complete the square around the original quadrilateral. The required area is the area of the square minus the area of the three right-angled triangles.
The required area is

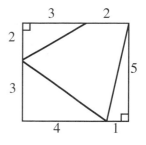

$$5 \times 5 - \frac{1}{2}(2 \times 3) - \frac{1}{2}(3 \times 4) - \frac{1}{2}(1 \times 5)$$

$$= 25 - 3 - 6 - 2.5$$

$$= 13.5.$$

Thus the area of the quadrilateral is 13.5 square units.

15. *Solution 1*

Since the dimensions are integers we can obtain them by examining the factors of the areas of floor and walls.
Area of the floor is $195 = 15 \times 13$ m^2.
Area of one wall is $104 = 8 \times 13$ m^2.
Area of the other wall is $120 = 8 \times 15$ m^2.

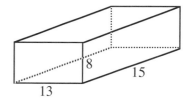

From the diagram, we see that the dimensions are 15 m by 13 m by 8 m.
The volume of the room is $15 \times 13 \times 8 = 1560$ m^3.

Solution 2

Let the length, width and height of the room, in metres, be l, w and h, respectively.
Then, $lw = 195$
$$lh = 120$$
$$wh = 104.$$

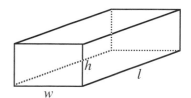

We don't need the individual dimensions; we only need the volume, which is $l \times w \times h$.
Suppose we just multiply the three equations together.

$$(lw)(lh)(wh) = 195 \times 120 \times 104$$

$$(lwh)^2 = 2\ 433\ 600.$$

$$lwh = \sqrt{2\ 433\ 600}$$

$$= 1560.$$

The volume of the room is 1560 m^3.

16.   Since the inner square has perimeter 20, its side length is 5 and its area
      is 25.
      Similarly the side length of the outer square is 7 and its area is 49.
      The shaded area is $49 - 25 = 24$.

17.   Each side of square Y is 15 cm and each
      side of square P is 8 cm.
      The hypotenuse of right-angled triangle T is
      $\sqrt{15^2 + 8^2} = \sqrt{289} = 17$ cm.
      The perimeter of T is $15 + 8 + 17 = 40$ cm.

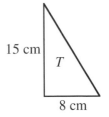

18.   We start by drawing perpendiculars from P
      and Q to meet SR at X and Y, respectively.
      Note that $XY = PQ = 18$ cm and that
      $$SX = YR = \frac{32 - 18}{2} = 7 \text{ cm.}$$
      By applying the Pythagorean Theorem in
      $\triangle PXS$ we find,

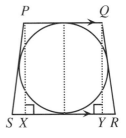

$$(PX)^2 + 7^2 = 25^2$$
$$(PX)^2 = 576$$
$$PX = 24.$$

      The diameter of the circle is 24 cm.

Chapter 5   Patterns
Exercises

1.   (a) 35   (b) 192   (c) $7 \times 720 = 5040$ (d) ✔

(e)   (f) $33 + 8 = 41$ (g)   (h) $34 + 55 = 89$

2.   (a) 6   (b) $2^{75} = 2^{72} \times 2^3 = (...6)(8) = ...8$

3.   Powers of 3 end in 3, 9, 7, 1, 3, 9, 7, 1, ... So all powers of 3 with a multiple of 4 in the exponent, such as 3^4, 3^8, 3^{12}, etc., will have 1 as the units digit.

(a) $3^{25} = 3^{24} \times 3 = (...1) \times 3 = ...3$ (b) $3^{75} = 3^{72} \times 3^3 = ...7$

4. $9^{55} = 3^{55} \times 3^{55} = 3^{110} = 3^{108} \times 3^2 = ...9$. The units digit is 9.

5.   Since $7^1 = 7$, $7^2 = 49$, $7^3 = ...3$, $7^4 = ...1$, $7^5 = ...7$, etc., then
$7^{100} = \left(7^4\right)^{25} = ...1$. The units digit is 1.

6.   Since every third bead is blue, the 39th bead is blue and the 40th bead is red.

7.   The pattern has six figures that repeat in it. Since 210 is divisible by 6, then the 210th figure will be ▲. Therefore the 211th figure will be ◇.

10. $37^2 = 1369$

11.   Starting from the upper right, the number of squares in each part of the diagram is 1, 3, 5, 7, 9, etc.

12. $S_1 = \frac{1}{2}$, $S_2 = \frac{1}{2} + \frac{1}{4} = \frac{3}{4}$, $S_3 = \frac{1}{2} + \frac{1}{4} + \frac{1}{8} = \frac{7}{8}$. We predict that the sum of all the terms will be $\frac{1023}{1024}$.

13. $(50-49)+(48-47)+\cdots+(2-1)=1+1+\cdots+1=25\times1=25.$

15.   (a) 136   (b) 1275   (c) 20 100

16.   (a) 66   (b) 2850   18. (a) 3240   (b) 7875

19. $50+51+\ldots+150=(1+2+\ldots+150)-(1+2+\ldots+49)$

$$=\frac{150\times151}{2}-\frac{49\times50}{2}$$

$$=11\,325-1225=10\,100.$$

20.   (a)   2 + 4 + 6 + ... + 56 + 58 + 60
                   60 + 58 + 56 + ... + 6 + 4 + 2

Add: $\overline{\text{62 + 62 + 62 + ... + 62 + 62 + 62}}$

The sum is $\dfrac{30\times62}{2}=930$.

(b) $2(1+2+\ldots+30)=2\left(\dfrac{30\times31}{2}\right)=930$

21.   (a)   1 + 3 + 5 + ... + 55 + 57 + 59
                   59 + 57 + 55 + ... + 5 + 3 + 1

Add: $\overline{\text{60 + 60 + 60 + ... + 60 + 60 + 60}}$

The sum is $\dfrac{1}{2}(30\times60)=30^2=900$.

(b) $(1+2+3+\cdots+58+59+60)-(2+4+6+\cdots+56+58+60)$

$$=1830-2(1+2+3+\cdots+28+29+30)$$

$$=1830-2(465)=900.$$

22.   (a) 1, 3, 6, 10   (b) 15
      (c) $1+2+\cdots+6=21,\ 1+2+\cdots+7=28,\ 1+2+3+\cdots+39+40=820$
      (d) The area of the nth figure is the sum of the first n positive integers.

23. (a) 3, 8, 13, 18, 23, 28   (b) 5   (c) The coefficient of n is 5.

24. (a) The difference is 8.   (b) 11, 19, 27, 35, 43, 51

26. (a) $t_n = 2n - 1$ (b) $t_n = 3n - 2$ (c) $t_n = 5n + 7$

(d) $t_n = 7n - 5$ (e) $t_n = n^2$

27. (a) The 4th figure has 16 triangles; the 5th figure has 25 triangles.

(b) 16, 25   (c) 36, 49, 900, $100^2 = 10\,000$ (d) n^2

(e) The sum of n consecutive odd integers is also n^2.

Explorations: 1. (b) 11, 30   2. (b) 12, 50, n (c) 2^{11}, 2^{19}, 2^{n-1}

Solutions to Problems

1. $999 \times 888\,888 = 887\,999\,112$

2. The number of squares needed for Figures 1, 2, 3, and 4 are 4, 7, 10, and 13, respectively. The number of squares in each figure is of the form $3n + 1$. Therefore the 40th figure would need $3(40) + 1 = 121$ squares.

3. There are 5 different cards in the pattern so every fifth one will be a soccer card. The 35th card will be a soccer card, the 36th will be a hockey card and the 37th will be a baseball card.

4. $490 - 491 + 492 - 493 + \ldots + 508 - 509 + 510$

$= (490 - 491) + (492 - 493) + \ldots + (508 - 509) + 510$

$= (-1) \quad + \quad (-1) \quad + \ldots + \quad (-1) \quad + \quad 510$

$= -10 + 510$

$= 500.$

5. The pattern of beads is *RG RRGG RRRGGG RRRRGGGG ...* .

The number of beads is $(1+1)+(2+2)+(3+3)+(4+4)+...$.

Since $1+2+3+...+9 = 45$, then, when we have 9 red and 9 green beads, there will be 90 beads altogether, of which 45 are red.

The next 10 beads in the pattern will be red, giving a total of 55 red beads in the first 100.

6. The number of integers in the sequence is $1+2+3+4+5+...$.

Note that $1+2+3+...+12+13 = \dfrac{13 \times 14}{2} = 91$.

Therefore the 100th integer in the sequence will be in the 14th group of integers. Since the 15th group of integers will be fives, the 14th group of integers will be fours and so the 100th integer is a four.

7. We add up the beads by adding

$(1+1)+(3+3)+(5+5)+(7+7)+...+(13+13)+...$

$= 2(1+3+5+7+...+13+...)$.

Since the sum of the first 7 odd positive integers is $7^2 = 49$, then there will be 49 red beads in the first 98 beads. (This will get us to a group of 13 red and 13 green beads). The next two beads will both be red giving a total of 51 red beads in the first 100 beads.

8. Every fifth symbol is O. Then the 210th symbol will be O and the 214th symbol will be ▲ .

9.

Number of layers	2	3	4	...
Number of cubes	$1+4$	$1+4+9$	$1+4+9+16$	...

So the number of cubes in all the completed structures is

$(1+4)+(1+4+9)+(1+4+9+16)+(1+4+9+16+25)+...$ and this sum must not exceed 1000.

If we build a total of eight structures we will use

$5+14+30+55+91+140+204+285 = 844$ cubes.

Then there will be 156 cubes left over (not enough to build a 9th structure).

10. $5 + 10 + 15 + 20 + \dots + 990 + 995 + 1000$

$$= 5(1 + 2 + 3 + \dots + 198 + 199 + 200)$$
$$= 5 \times \frac{200 \times 201}{2}$$
$$= 100\,500.$$

11. The difference between successive numbers in the pattern is 3.
So each term has form $3n + k$. For $t_1 = 2$, k must be -1. Therefore $t_n = 3n - 1$, where $n = 1, 2, 3, \dots$.
Therefore $t_{100} = 3(100) - 1 = 299$.

12. Since the difference between successive terms is 4 and the first term is 3, then $t_n = 4n - 1$, where $n = 1, 2, 3, \dots$. We want to find n where $4n - 1 = 439$.
The value of n is 110, so there are 110 numbers in the pattern.

13. The sum is $2(1 + 2 + 3 + \dots + 14) + 15$

$$= 2 \times \frac{14 \times 15}{2} + 15$$
$$= 210 + 15$$
$$= 225.$$

14. $3 + 6 + 9 + 12 + 15 + \dots + 99$

$$= 3(1 + 2 + 3 + 4 + 5 + \dots + 33)$$
$$= 3 \times \frac{33 \times 34}{2}$$
$$= 1683.$$

15. In the table, the numbers in the odd numbered rows are increasing and in column S these numbers are $4, 12, 20, \dots$. That is, they increase by 8. Thus 100 will be in column S and 101 will also be in column S, directly below 100.

16. (a) We could just continue the pattern and we would find the 7th and
          8th rows are

	22	23	24	25	26	27	28	
and	29	30	31	32	33	34	35	36.

So the number directly under 25 is 32. Can we do this without
actually writing down the rows?

If we wish to avoid writing line after line, we need a pattern to help
us find either the last number in any line or the first number in a
line.

Since the number of integers used is one in line 1, two in line 2,
three in line 3, and so on, the last integer in line n is

$1+2+3+...+n = \dfrac{n(n+1)}{2}$, a formula we already know. For

example, the last integer in line 7 is $\dfrac{7 \times 8}{2} = 28$, as we have already

seen.

Even more easily, the first numbers in the rows are 1, 2, 4, 7, 11,
16, 22, 29, ... Here there is a simple pattern of differences. Using
this we can say that every number in the row beginning with 29 is
7 greater than the number above it, so 25 has 32 below it.

(b) Using the first number pattern, 60 is below 50, 87 is below 75, and
     114 is below 100.

17. (a) Let us tabulate powers of 2 and powers of 3

$$2^1 = 2 \qquad 3^1 = 3$$
$$2^2 = 4 \qquad 3^2 = 9$$
$$2^3 = 8 \qquad 3^3 = 27$$
$$2^4 = 16 \qquad 3^4 = 81$$
$$2^5 = 32 \qquad 3^5 = 243$$

The last digits in powers of 2 form a repeating sequence of four
digits, namely, 2, 4, 8, 6, 2, 4, 8, 6, 2, ...

Similarly, the last digits in powers of 3 form the repeating sequence of four digits 3, 9, 7, 1, 3, 9, 7, 1, 3, ...

Therefore $2^7 + 3^7$ is the sum of an integer ending in 8 and an integer ending in 7. This produces an integer ending in 5.
Of course, we could have directly evaluated the sum:

$$2^7 + 3^7 = 128 + 2187 = 2315 \text{ which ends in a 5.}$$

(b)   The discussion in (a) enables us to quickly solve the problem in (b). Since 32 is divisible by 4 we know that 2^{32} will end in 6. It follows that 2^{34} will end in 4. Similarly 3^{32} will end in 1 and so 3^{34} will end in 9.
Therefore $2^{34} + 3^{34}$ will end in 3.

Chapter 6   *Algebra*
Exercises

2.   (a) $8n+20$ (b) $12a+21$ (c) $3n+5$ (d) $a+2$

3.   (a)   The answer is always 2.
      (b)   The answer is the same as the number picked.

5.   If $p=3$, $(p-1)(4p-2)=(3-1)(12-2)=(2)(10)=20$.

6. $\dfrac{(12+2)(12-2)}{4-2}=\dfrac{140}{2}=70$

7.   (a)   From the first balance we get $3\,\square = 2\,\blacktriangle + 2\,\bigcirc$.
             From the second balance we get $3\,\square = 6\,\bigcirc$.
             From these two equations we get $6\,\bigcirc = 2\,\blacktriangle + 2\,\bigcirc$.
             Therefore $2\,\blacktriangle = 4\,\bigcirc$ and so $1\,\blacktriangle = 2\,\bigcirc$.
             Thus two circles will balance one triangle.

      (b)   From the first balance we get $3\,\square = 1\,\square + 2\,\bigcirc + 4\,\blacktriangle$.
             But the second balance tells us that $2\,\bigcirc = 4\,\blacktriangle$.
             Therefore $3\,\square = 1\,\square + 4\,\blacktriangle + 4\,\blacktriangle$.
                       $2\,\square = 8\,\blacktriangle$.
             Therefore four triangles will balance one square.

9.   (a) $12x$ (b) $10m$ (c) $5t+13$ (d) $9a+9$
      (e) $4m+22$

10.   (a) 8   (b) 9   (c) 20   (d) 15

11.   Let the number be n.   12. Let the number be x.
                $8n+4=100$ $7x-6=85$
                $8n=96$ $7x=91$
                $n=12$. $x=13$.
      The original number is 12.   The original number is 13.
      Note: Exercises 11 and 12 can easily be done without the use of algebra.

13.　*Solution 1*

　　Let the smallest integer be n. Then the other integers are $n+2$ and $n+4$.

$$n+n+2+n+4 = 57$$
$$3n+6 = 57$$
$$3n = 51$$
$$n = 17.$$

　　The numbers are 17, 19 and 21.

　　Solution 2

　　Here's an interesting fact. If you add an odd number of consecutive integers, their sum divided by how many numbers there are gives the middle number. In other words, the average of the numbers is the middle number. This is also true for consecutive odd or consecutive even integers. Let's check this.

　　$5+6+7+8+9 = 35$ and $35 \div 5 = 7$.

　　$10+12+14 = 36$ and $36 \div 3 = 12$.

　　Convince yourself that this is *not* true for an even number of consecutive integers.

　　Using this fact, this problem can be done as follows:

　　The average of the three numbers is $57 \div 3 = 19$.

　　The numbers are $19-2 = 17$, 19, and $19+2 = 21$.

14.　The average of the five integers is $190 \div 5 = 38$.

　　Thus 38 is the middle integer and so the five integers are 36, 37, 38, 39, and 40.

15.　Let the first integer be y.

$$y+(y+2)+(y+4)+(y+6) = 1268$$
$$4y+12 = 1268$$
$$4y = 1256$$
$$y = 314.$$

　　The four integers are 314, 316, 318, and 320.

16. (a) $5n + 28 = 6n$ (b) $7y - 9 = 2y + 46$

              $n = 28$. $5y = 55$

                                        $y = 11$.

17. Let the number of houses be n.

Then the number of apartments is $n + 16$.

$$n + (n + 16) = 80$$
$$2n = 64$$
$$n = 32.$$

Petra delivers papers to 32 houses.

18. Let b represent the number of birds and let p represent the number of puppies.

$$b + p = 30$$
$$2b + 4p = 94.$$

Solution 1

Using a chart (with $b + p = 30$).

b	p	$2b + 4p$
15	15	$2(15) + 4(15) = 90$
16	14	$3(16) + 4(14) = 32 + 56 = 88$
14	16	$2(14) + 4(16) = 92$
13	17	$2(13) + 4(17) = 94$

Therefore there are 13 birds and 17 puppies in the store.

Solution 2

Using algebra:

$b + p = 30$ or $b = 30 - p$

$$2b + 4p = 94$$

$$2(30 - p) + 4p = 94$$

$$60 - 2p + 4p = 94$$

$$60 + 2p = 94$$

$$2p = 34$$

$$p = 17.$$

$$b = 30 - 17$$

$$= 13.$$

Therefore there are 13 birds and 17 puppies in the store.

Solutions to Problems

1. Suppose Pat was paid x dollars for the first day. Then she was paid successively $x + 5$, $x + 10$, $x + 15$, and $x + 20$ dollars for the next four days.

$$x + (x + 5) + (x + 10) + (x + 15) + (x + 20) = 150$$

$$5x + 50 = 150$$

$$5x = 100$$

$$x = 20.$$

Therefore Pat was paid $40 on the last day.

2. The cost of the light is $370 - \$330 = \$40$.

    The cost of the stand is $\frac{1}{4}(\$40) = \$10$.

    Therefore the cost of the bicycle is $330 - \$10 = \$320$.

3. Before adding six, the doubled number was $28 - 6 = 22$.

    Then the number was 11.

    First adding 6 and then doubling would give $(11 + 6) \times 2 = 34$.

4.   If three times a number is 36, the number is 12.
     Then four times the number is 48.

5.   A soft drink costs $\$2.80 - \$2.05 = \$0.75$.
     A burger costs $\$3.30 - \$2.05 = \$1.25$.
     Therefore an order of fries costs $\$2.05 - \$1.25 = \$0.80$.
     A burger, two orders of fries and a drink costs
     $\$1.25 + 2(\$0.80) + \$0.75 = \$3.60$.

6.   *Solution 1*
     Since the sum of the column sums must equal the sum of the row sums,
     $$T + 19 + 20 + 30 = 28 + 30 + 20 + 16$$
     $$T = 25.$$

     Solution 2
     From the first row we get $4D = 28$,
     so $D = 7$.
     From the second row we get
     $2D + 2N = 30$

     $$2N = 16$$
     $$N = 8.$$
     From the third row,
     $8 + W + P + 7 = 20$
     $\quad W + P = 5$.
     From the fourth row, $P + P + W + 8 = 16$

     $$P + 5 + 8 = 16$$
     $$P = 3.$$
     Therefore $T = 7 + 7 + 8 + 3 = 25$.

				Row Sum
D	D	D	D	28
D	D	N	N	30
N	W	P	D	20
P	P	W	N	16
Column Sum T	19	20	30	

7.   (a) $n(n+2)$ is odd if n is odd and even if n is even.
     (b) Since n and $n+1$ are consecutive integers, one of them will be
          even and so $n(n+1)$ is always even. Thus $n(n+1)$ is always
          divisible by 2.
     (c) Since $n + n = 2n$, it is always divisible by 2.
     (d) n^2 is even if n is even and odd if n is odd.

8.   The answer is always 3.

9.   If Janine scored r two-point baskets and t three-point baskets, then
      $2r + 3t = 36$. The possible combinations, (r, t), that she could have
      scored are $(0, 12)$, $(3, 10)$, $(6, 8)$, $(9, 6)$, $(12, 4)$, $(15, 2)$, and $(18, 0)$.

10.   We could divide the 60 students into five groups of 12 with two of the
      groups being boys and three of the groups being girls.
      Then there are 24 boys and 36 girls.

11.   Let the amount left to each of the grandchildren be x dollars.
      Then the son is left $8x$ dollars and the widow is left $(x + x + 8x + 20\,000)$
      dollars.
      Hence $x + x + 8x + (x + x + 8x + 20\,000) = 120\,000$

$$20x + 20\,000 = 120\,000$$
$$20x = 100\,000$$
$$x = 5000.$$

      Each grandchild receives $5000, the son receives $40\,000, and the widow
      receives $70\,000.

12.   Since $2n$ is always even, then $2n + 1$ is always odd.
      Of the others, n^2 is even if n is even, $\sqrt{n}$ is even if n is an even perfect
      square such as 36, and $\frac{n}{2}$ is even whenever n is a multiple of 4.

13.   The sum of each row, column, and diagonal
      is $16 + 10 + 4 = 30$.
      From the first row, $16 + 2 + A = 30$
      $$A = 12.$$
      Using the diagonal, $12 + 10 + B = 30$
      $$B = 8.$$
      From the first column, $16 + C + 8 = 30$
      $$C = 6.$$
      Thus $A = 12$, $B = 8$ and $C = 6$.

16	2	A
C	10	E
B	D	4

14. Since half the number is 50, the number is 100.

15. $6 \triangle 2 = 6 \times 2 + \frac{6}{2} = 12 + 3 = 15$.

16. Let r represent the number of red marbles.
    Then Fran purchased $2r$ white marbles and $3r$ blue marbles.
    It follows that $r + 2r + 3r = 150$
    $$6r = 150$$
    $$r = 25.$$
    Fran purchased 25 red marbles.

17. All 20 frames need at least 2 wheels.
    This uses up 40 wheels with 8 left over.
    Since the 8 must be used, they would go on tricycles, making 8 tricycles and 12 bicycles.

18. $7 \triangle 5 = \dfrac{7 + 5}{3} = 3$.

    $3 \triangle 1 = \dfrac{3 + 1}{4} = 1$.

    Therefore $(7 \triangle 5) \triangle 1 = 1$.

19. The 'magic' sum is $8 + 9 + 4 = 21$, so the centre square contains a 7. Then the square on the lower right has 6 in it giving $4 + n + 6 = 21$.
    Therefore $n = 11$.

8		
9		5
4	n	

20. If Juan won 3 games then Mary lost 3 points so that she must have had 8 points before losing in order to have a final total of 5.
    If Mary had 8 points before losing then she must have won 4 games.
    If Mary won 4 games and Juan won 3 games, they played a total of 7 games.

21. Since R is on a main diagonal and the numbers 1, 2 and 3 have already been used on this diagonal, then $R = 4$.

    The easiest way to look at how to arrange the numbers is to look at boxes P and Q. Q must be either 2 or 3 but since there is already a 2 in the same column as Q, we conclude that $Q = 3$ and $P = 2$.

1			
	2		
		3	
2	3	1	4

'cannot be 2'

From this point we simply fill in the boxes according to the rule that each row, column and diagonal contains each of the numbers 1, 2, 3, and 4.

Doing this, we arrive at the arrangement of numbers shown.

1	4	2	3
3	2	4	1
4	1	3	2
2	3	1	4

We see that $K = 1$ and $N = 2$.

Note: It is not necessary that we complete all the boxes but it is a useful way to verify the overall correctness of our work.

Chapter 7   *Counting and Probability*
Exercises

1. 15   2.   161

3. (a)   15
    (b)   25. We usually are averaging numbers that are distributed evenly.
          When there are a few that differ greatly from the others, averages
          can become somewhat meaningless, because we think of average
          as indicating the middle, more or less.

4. 12   5. 40   6. 24   7. 12

8. $\frac{1}{18}, \frac{17}{18}$ 9. $\frac{1}{9}$ 10. $\frac{3}{10}, \frac{11}{20}$ 11. $\frac{3}{5}$

12. (a) 7 (b) 12

13. A total of 19 three-digit numbers can be formed. These numbers are
     666, 663, 636, 366, 668, 686, 866, 336, 363, 633, 338, 383, 833, 638,
     683, 368, 386, 836, and 863.

14. Let the three kinds of chocolate bars be *A*, *B* and *C*. The ten possible
     selections are *AAA*, *BBB*, *CCC*, *AAB*, *AAC*, *BBA*, *BBC*, *CCA*, *CCB*,
     and *ABC*.

Solutions to Problems

1. If the average of the four numbers is 5, then their sum is $4 \times 5 = 20$.
    Since two of the numbers are 3 and 4, the sum of the other two is 13.

2. If the sum of three numbers is 180, their average is 60.
    The first number is 70 and the second is 56, so the third number is
    $180 - 70 - 56 = 54$.
    [Note that since the first is 10 more than the average and the second is 4
    less than the average, then the third must be 6 less than the average, or
    54.]

3.   The sum of the numbers is $26 + 28 + 29 + 38 + 69 = 190$.

       The average of the numbers is $190 \div 5 = 38$.

       [Could you do this problem by simply examining the numbers?]

4.   There are $6912 \div 8 = 864$ pumpkin plants.

       On average there are $864 \div 12 = 72$ plants per hectare.

5.   Reading from the graph, Jean's five test marks are 80, 70, 60, 90, and 80.

       Her average mark is $\frac{80+70+60+90+80}{5} = \frac{380}{5} = 76$.

6.   There are 30 marbles and 5 of them are yellow.

       The probability of selecting a yellow marble is $\frac{5}{30} = \frac{1}{6}$.

7.   (a)   We can add these numbers, obtaining 165, so the average is
             $165 \div 11 = 15$.
             An easier approach is to note that the average of 10 and 20 is 15,
             the average of 11 and 19 is 15, and so on for every similar pair.
             This will always work if the number of terms in the sum is odd.

       (b)   There are 51 numbers in the set, so the average is $\frac{50+100}{2} = 75$.

       (c)   Here we have an even number in the set. It is still true that
             $10 + 21 = 11 + 20 = 12 + 19$, and so on. The average is $\frac{10+21}{2} = 15\frac{1}{2}$.

8.   If seven consecutive numbers have an average of 20, they must be 17, 18,
       19, 20, 21, 22, 23. (Starting with 20 we go up three numbers, then down
       three numbers.)

9.   The sum of the fifteen numbers is $5 \times 22 + 10 \times 30 = 410$.

       The average of the fifteen numbers is $410 \div 15 \doteq 27.3$.

10.   If five numbers have an average of 12, their sum is $12 \times 5 = 60$.

       If six numbers have an average of 14, their sum is $14 \times 6 = 84$.

       The sixth number is $84 - 60 = 24$.

11.   If the team average for five games is 27, then the number of points
       scored is $27 \times 5 = 135$.
       In the fifth game Gainesville scored $135 - 97 = 38$ points.

12.   In the first four games Joan scored $4 \times 12.5 = 50$ points.
       In the next five games Joan scored $5 \times 15 = 75$ points.
       In the ten games Joan scored $50 + 75 + 20 = 145$ points, so her average
       for each game was 14.5 points.

13.   There are four sets of three letters: MAT, MAH, ATH, MTH.
       For any one of these there are 3 choices for the first, 2 choices for the
       second, and 1 for the third, a total of 6 different arrangements.
       Hence there are $4 \times 6 = 24$ arrangements of three letters.
       [Note that for the letters M, A and T we have MAT, MTA, AMT, ATM,
       TAM, TMA.]
       Alternatively we can say that there are 4 choices for the first letter, 3
       choices for the second letter, and 2 choices for the third letter. Therefore
       there are $4 \times 3 \times 2 = 24$ possible arrangements.

14.   For pages 1 to 9 we require $1 \times 9 = 9$ digits.
       For pages 10 to 99 we require $2 \times 90 = 180$ digits.
       From pages 100 to 300 there are 201 pages so we require $3 \times 201 = 603$
       digits.
       In total we require $9 + 180 + 603 = 792$ digits.

15.   There are 5 choices for the first person in the line, and then 4, 3, 2, and 1
       choices for the remaining four people. Therefore there are
       $5 \times 4 \times 3 \times 2 \times 1 = 120$ possible arrangements.

16.   Since two students have neither, there are $35 - 2 = 33$ who have roller
       blades, skateboards, or both.
       Since there are $24 + 19 = 43$ who have one or the other, there must be
       $43 - 33 = 10$ who have both.

17. There are 60 numbers. Of these, the ones containing 4 are 4, 14, 24, 34, 54, and 40, 41, ..., 49, and there are 15 of these.

The probability of choosing a number containing 4 is $\frac{15}{60} = \frac{1}{4}$.

18. We consider cases:

If Bob uses two 6's and either 5 or 7 he can form two numbers: 665, 667.

If Bob uses two 5's and either 6 or 8 he can form four numbers: 565, 655, 585, 855.

If Bob uses two 5's and one 7, he can form three numbers: 557, 575, 755.

If Bob uses three of 5, 6, 7, 8, he has 2 choices for the ones digit, 3 choices for the hundreds digit, and 2 choices for the tens digit, so he can form 12 numbers.

Therefore he can form $2 + 4 + 3 + 12 = 21$ odd numbers.

19. If one card is selected from each deck, there are $12 \times 12 = 144$ possible pairs of cards.

He can obtain a sum of 15 by any of the following pairs:

Red	3	4	5	6	7	8	9	10	11	12
Blue	12	11	10	9	8	7	6	5	4	3

Since there is a total of 10 pairs, the probability that the cards sum to 15 is $\frac{10}{144} = \frac{5}{72}$.

20. (a) If $a + b = 25$ and a and b are positive integers, there are 24 possible values for a: 1, 2, 3, ..., 23, and 24. For each of these there is an acceptable value for b, (namely $25 - a$).
Hence there are 24 ordered pairs.

(b)   If $c = 1$, then $a + b = 25$ and there are 24 ordered triples, using the pairs from part (a) with $c = 1$.

If $c = 2$, then $a + b = 24$ and there are 23 ordered triples.

If $c = 3$, then $a + b = 23$, and there are 22 ordered triples.

... and so on ...

If $c = 23$, then $a + b = 3$, and there are 2 ordered triples.

If $c = 24$, then $a + b = 2$, and there is 1 ordered triple.

In total there are $24 + 23 + 22 + ... + 3 + 2 + 1 = \dfrac{24(24 + 1)}{2} = 300$

ordered pairs.

Chapter 8
Exercises
           Miscellaneous Problems

1.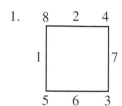

2. When $a+b+c+d=12$ and the sum of the integers on each side is k, we get
$$4k-36=12$$
$$4k=48$$
$$k=12.$$
The sum of the integers along each side is 12.

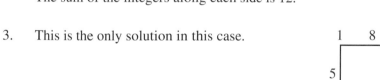

3. This is the only solution in this case.

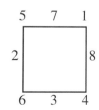

4. When $a+b+c+d=16$, we get $4k-36=16$
$$4k=52$$
$$k=13.$$
The sum of the integers on each side is 13.

5.

8	3	2
4		6
1	7	5

5	7	1
2		8
6	3	4

6. When the sum of the integers on the four corners is 12 or 24, we get one solution in each case. When the sum of the integers on the four corners is 16 or 20, we get two solutions in each case. Therefore there is a total of six possible solutions.

Solutions to Problems

1. In 45 days there are 6 full weeks and 3 extra days. There could be 7 Mondays in the first 45 days.

2. Since there are 30 days in November, the only days which can occur five times are those that fall on November 1 or 2. Since November 12 is a Wednesday, then November 5 is also a Wednesday. Therefore November 1 is a Saturday and November 2 is a Sunday. Each of these will occur five times.

3. If the mass of all people in the elevator is 100 kilograms over the limit, their total mass is 1600 kilograms. If their average mass is 80 kilograms, then there are $\frac{1600}{80} = 20$ people in the elevator.

4. A number is equal to the product of a divisor and the quotient, plus the remainder.
    The number is $6 \times 9 + 4 = 58$.

5. We use a chart to calculate the per minute costs:

Person	Cost	Time (minutes)	Cost per minute
John	$5	1	$5
David	$1	1	$1
Clare	$2	4	$0.50
Barbara	$5	4	$1.25
Sanjay	$1	5	$0.20

John's call cost the most per minute.

6.　In the ten levels there are

$$10+9+8+7+6+5+4+3+2+1 = \frac{10(10+1)}{2} = 55 \text{ blocks.}$$

7.　The volume of a popped kernel is $5 \times 1.6 = 8 \text{ cm}^3$. Then the length of a side of a piece of popped corn is 2 cm.

8.　On parts A and B Rami uses $1 \times 10 + 2 \times 10 = 30$ minutes. He now has 30 minutes for part C, and since there are five questions, he can spend an average of 6 minutes per question.

9.　If the snail climbs 4 m and slips back 2 m each day, it gains a total of 2 m daily. In eight days it climbs 16 m. On the next day it will climb 4 m and reach the top without slipping back. Therefore it will reach the top in nine days.

10.　Each side of the shaded square is $8 - 5 = 3$ cm.

Then the area of the entire figure is $2 \times 6 \times 8 + 3 \times 3 = 105 \text{ cm}^2$.

11.　If the ball bounces $\frac{2}{3}$ of the height from which it falls each time, after

the third bounce it will rise $\frac{2}{3} \times \frac{2}{3} \times \frac{2}{3} \times 60 = \frac{160}{9} = 17\frac{7}{9}$ cm or

approximately 17.8 cm.

We can illustrate the height reached as shown in the diagram.

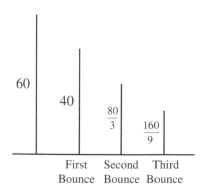

12. Suppose each of the cartons contains 100 mL. Then the carton of whipping cream contains 35 mL of butterfat and the carton of milk contains 2 mL of butterfat. The mixture contains 200 mL of which 37 mL is butterfat. Therefore the percentage of butterfat in the mixture is

$$\frac{37}{200} \times 100 = 18.5.$$

13. Perfect square products can be obtained as follows:

Yellow Card	1	1	3	5	9	9	12
Blue Card	1	4	3	5	1	4	3
Product	1	4	9	25	9	36	36

There are 7 winning pairs.

14. Number the points 1 to 10.
Point 1 can be connected to all 9 other points.
Point 2 can now be connected to 8 other points (but not to number 1 because that's already done).
Point 3 can now be connected to 7 other points.
Continuing the argument we finally get point 9 connected only to point 10.
There are $9+8+7+6+5+4+3+2+1 = 45$ possible chords.

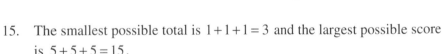

15. The smallest possible total is $1+1+1=3$ and the largest possible score is $5+5+5=15$.
Since the scores are all odd, the sum of any three of them will be odd. Hence the only possible scores are 3, 5, 7, 9, 11, 13, and 15, and all of these can be obtained.
They are $1+1+1$, $1+1+3$, $1+1+5$, $1+3+5$, $1+5+5$, $3+5+5$, and $5+5+5$.
There are seven possible scores.

16. If every bus has the same number of passengers, we need to find a number no larger than 49 that divides into 1161.
Now $1161 = 9 \times 129 = 27 \times 43$.
The largest number that divides into 1161 and is not larger than 49 is 43.
If there are 43 people on each bus, then 27 buses are required.

17. By actually tracing the routes on the diagram, we find there are 10. A difficulty in tracing is that we are never sure whether we got them all, unless we do it with extreme care. A better way is to observe that we have to use three east moves and two south moves. We can list all possible routes by using E for each of the east moves and S for each of the possible south moves. Then the routes are: EEESS, EESES, ESEES, SEEES, SEESE, SESEE, SSEEE, ESSEE, EESSE, ESESE.

18. Number the houses from 1 to 10.
The first house can be connected to 9 others.
The second house can be connected to 8 others.
(Is this sounding like the solution to problems 14? Absolutely!)
The number of connections needed is
$9 + 8 + 7 + 6 + 5 + 4 + 3 + 2 + 1 = 45$.

19. In $\frac{2}{5}$ of an hour, the minute hand would move $\frac{2}{5} \times 60 = 24$ minutes.
The correct time is 4:24.

20. It takes $\frac{1}{6}$ of the pitcher's volume for one full glass. If there is enough water to fill four glasses, the pitcher must be $\frac{4}{6}$ or $\frac{2}{3}$ full.

21. If Frances skates five laps, Sangeeta skates ten laps. If Sangeeta skates ten laps, Timoula skates forty laps.
The total number of laps is $5 + 10 + 40 = 55$.

22.   If the length of the cable is x metres, then

$$x^2 = 1000^2 + 750^2$$
$$= 1\,000\,000 + 562\,500$$
$$= 1\,562\,500.$$
$$x = \sqrt{1\,562\,500} = 1250.$$

The cable length is 1250 m, and if it takes 8 minutes to reach the top, the speed of the chair is $\frac{1250}{8} = 156.25$ metres per minute, or 156.3 metres per minute, correct to one decimal.

Note: It's always nice to avoid long calculations. You have noted in earlier questions that a triangle whose sides are 3, 4, and 5 is right-angled. Hence a triangle with sides 6, 8, and 10 is also right-angled, and in general a triangle with sides $3n$, $4n$ and $5n$ is right-angled. Here we have one side $3 \times 250 = 750$, a second side $4 \times 250 = 1000$, so the third side is $5 \times 250 = 1250$, and you can avoid the calculation!

23.   There are $185 + 110 = 295$ students who are wearing blue jeans, have blue eyes, or who have blue eyes and are wearing blue jeans. But there are only 250 students in total and 55 are not wearing blue jeans and do not have blue eyes. The number of students wearing blue jeans or having blue eyes is $250 - 55 = 195$. Therefore there are $295 - 195 = 100$ students who have blue eyes and are wearing blue jeans.

24.   In both cases the family travelled at 90 km/h for the first 600 km, so the time required was the same. In the first case the last 200 km was travelled at 100 km/h, so the time required was $\frac{200}{100} = 2$ hours. In the second case the last 200 km was travelled at 90 km/h, so the time required was $\frac{200}{90} = 2\frac{2}{9}$ hours. The trip would have taken $\frac{2}{9}$ hours, or about 13 minutes longer if they had travelled at 90 km/h for the entire journey.

25. If there are twenty-four different numbers, there must be six which have a ones digit 1, six which have a ones digit 2, six which have a ones digit 3, and six which have a ones digit 4. The sum of the ones digit column is $6+12+18+24=60$. It is also true that in the tens, hundreds, and thousands column there are six 1's, 2's, 3's, and 4's. The sum of the tens column is $10 \times 60 = 600$; the sum of the hundreds column is $100 \times 60 = 6000$; the sum of the thousands column is $1000 \times 60 = 60\,000$.

The total sum is $60\,000 + 6000 + 600 + 60 = 66\,660$.

26. Starting from A the shortest distance is to H, a distance of 4 kilometres, so place $\boxed{4}$ beside H. Next place $\boxed{7}$ beside F. Since $\boxed{7} + 1 = 8$ is the minimum possible, put $\boxed{8}$ beside S. Following the procedure of always looking for the smallest possible, next put $\boxed{10}$ beside R, $\boxed{11}$ beside M, $\boxed{14}$ beside each of D and G, and finally $\boxed{18}$ beside B.

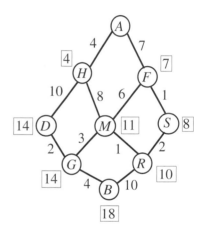

The shortest route from Amy's Cove to Burley is 18 kilometres, passing through Fractionville, Steamroller, Rowntree, Manhill, Greenlee, and ending at Burley. Note that Hillsdale and Deepwell are missed.
If John is delivering mail and must pass through Hillsdale and Deepwell, returning to Amy's Cove, his best route is to take this route to Burley, then to go back to Greenlee, Deepwell, Hillsdale, and finally to Amy's Cove. This adds $4 + 2 + 10 + 4 = 20$ kilometres, making the total distance 38 kilometres.

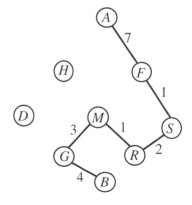

27. Each of the points A, B, C, D, E, F, G, and H, gives a right-angled triangle with XY. Also the points J and K give right-angled triangles because, using the Pythagorean Theorem, $XJ = \sqrt{2}$, $JY = \sqrt{2}$ and $XY = 2$. There are ten possible positions for Z.

```
  •   A   •   E   •
      •       •
  •   B   J   F   •
      •   •   •
  •   X   •   Y   •
      •       •
  •   C   K   G   •
      •   •   •
  •   D   •   H   •
      •   •   •
```

28. It's interesting that we do not require the actual dimensions to do this problem.
If the width is w units, then the length is $2w$ units. In going from R to Q the caterpillar travels $4w$ units. In going from P to R and then on to Q the spider travels $7w$ units.

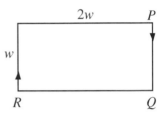

The spider's speed must be $\frac{7}{4}$ that of the caterpillar, or $\frac{7}{4} \times 2 = 3.5\,\text{mm/s}$.

29. The sum of the numbers 1, 2, 3, ..., 9 is 45.
The sum along each edge must be the same as that along the other edges, so the edge sum must be a multiple of 4.
The only multiples of 4 which allow for a number 1 to 9 to be in the middle are 44, with 1 in the middle, 40, with 5 in the middle, and 36, with 9 in the middle.
There are three entries possible for the middle circle.

30. If the sum of nine consecutive integers is 99, the average (or the middle number) is 11. There are four numbers below 11 and four numbers above 11.
Then the largest of the nine numbers is $11 + 4 = 15$.

31.   A red edge is contained by two faces.
Hence there must be at least three red
edges, because there are six faces.
The diagram shows how this can be done.

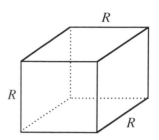

32.   If the matchsticks marked *a*, *b* and *c* are moved as in the diagram, there
are three squares rather than four.

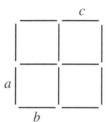 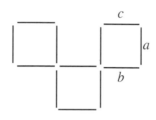

33.   This is a marvellous problem. When I was a young student I spent hours
trying to do this. I could always get eight connections, but never the
final one. Then I discovered, years later, that it is impossible to do.
Here's why:

We connect *B* to *E*, *B* to *G*, and *B* to *W*,
as shown.

Next we connect *A* to *E*, *A* to *G*, and
A to *W*, as shown, using either route 1
or route 2 to connect *A* to *W*. If route
1 is used, then *C* cannot be connected
to *G*. If route 2 is used, then *C*
cannot be connected to *E*. It is
impossible to make all connections.
This is an amazing fact.

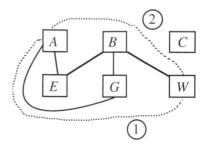

34.   If 70% of the class can swim, there are $\frac{70}{100} \times 380 = 266$ swimmers.

Since 200 girls can swim then there are 66 boys who can swim.
Suppose the number of boys is b.
Since 60% of the boys can swim, then $0.6b = 66$

$$b = \frac{66}{0.6} = 110.$$

There are $380 - 110 = 270$ girls.

35.   (a)   Since the bottom is 4 cm by 4 cm, we can fit four cubes with their bottom faces on the base of the big cube, and then another four cubes on top of these.
We can fit eight 2 cm cubes into the box.
[Note that this can also be done by using the fact that the big cube has volume $4 \times 4 \times 4 = 64$ cm^3 and each of the small ones has volume $2 \times 2 \times 2 = 8$ cm^3.]

(b)   We can fit nine 2 cm cubes on the bottom of the 7 cm cube, as shown. We can stack the cubes three high so we can put $3 \times 9 = 27$ of the smaller cubes in the 7 cm cube.
[Note that we cannot do this packing by using volumes as in (a).]

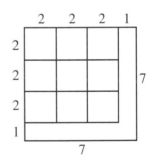

36.   Many problems are easier to do if we re-interpret the question. This is one such, for it is easier to ask "How many have exactly two red faces?", and if there are four white faces then there must be two red faces.

If the larger cube has twenty-seven smaller cubes, then it has dimensions $3 \times 3 \times 3$. A small cube at a vertex will have three red faces, and a small cube in the middle of a face will have one red face. This means that a small cube lying on an edge and in the middle of that edge will have two red faces, as shown. There are four of these in each of the six faces, but each appears in two faces, so there are 12 in total.

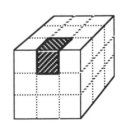

37. *Solution 1*

    If the number on the blue die decreases with each roll, its smallest possible value on the fourth roll is 1, so on the second roll it could be no smaller than 3. If the number on the yellow die increases with each roll, it must be at least 4 on the fourth roll, so on the second roll it must be at least 2. The smallest possible sum is 5.

    Solution 2

    If the blue die decreases with each of four rolls, the smallest set of values for the four rolls is 4, **3**, 2, and 1. If the yellow die increases with each of four rolls, the smallest set of values is 1, **2**, 3, and 4. The smallest sum on the second roll is $3 + 2 = 5$.

38. (a) It takes 5 lines for GAUSS to be written again.
    Hence GAUSS appears on lines 1, 6, 11, 16, 21, ...
    It takes 4 lines for 1234 to be written again.
    Hence 1234 appears on lines 1, 5, 9, 13, 17, 21, ...
    GAUSS 1234 will appear on line 21.
    Note that the Lowest Common Multiple of 5 and 4 is 20, so GAUSS 1234 will appear every 20 lines; that is, on lines 1, 21, 41, 61, ...

    (b) GEOMETRY has 8 letters and 123456 has 6 digits.
    The Lowest Common Multiple of 8 and 6 is 24, so GEOMETRY 123456 appears every 24 lines; that is, on lines 1, 25, 49, 73, ...
    The third appearance is on line 49.

39. If the sides of the triangle are 30, 40 and 50, then the triangle is a right-angled triangle, because the side lengths are multiples of 3, 4 and 5. The sides about the right angle have lengths 30 and 40.

    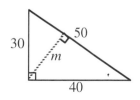

    Using 30 as the altitude, the area of the triangle is $\frac{1}{2} \times 40 \times 30 = 600$.

    Using m as the altitude, the area of the triangle $\frac{1}{2} \times 50 \times m = 25m$.

    Therefore $25m = 600$

    $\qquad\qquad m = 24$.

    The shortest altitude is 24.

40. Steven arrived at 10:20 a.m.

At 80 km/h, it takes him $\frac{60}{80} \times 60 = 45$ minutes to go 60 kilometres.

He left home at 25 minutes before 10 a.m. or at 9:35 a.m.

41. Each of the four triangular pieces is an
isosceles triangle with a right angle,
and base 8. Together, the four
triangles make up a square with side 8.
The area of this smaller square is 64.
The area of the original square is 100.
The area of the cross is 36.

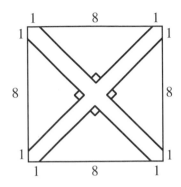

42. The area of the smallest tile is 18.
The area of the next smallest tile is 36.
The area of the largest tile is 54.
The sum of the three areas is 108.
The area of the square pattern created is a multiple of 108, so let it be
108m.

We want m to make 108m a square. Since $108 = 2 \times 2 \times 3 \times 3 \times 3$, then m
must equal 3 so that each prime factor occurs an even number of times.
Then the square pattern has area $3 \times 108 = 324$, which is a square with
dimensions 18×18.
Therefore we need 9 stones to form the smallest possible square.
This can be done is many ways; you may want to convince yourself by
drawing a layout.

43. This is possible as noted in the diagram.

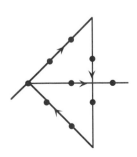

44. (a) There is more than one way of doing this. Here are two solutions:

     4   1
    8 1   3 6
   3 9   7 8
  5 7 2 6   9 4 2 5

(b) For a sum of 17 along each side, we get

              1
           5 7
         9 6
       2 8 4 3

(c) If the sum along each side is k, then
$$a+d+e+b+b+f+g+c+c+h+i+a=3k.$$
Since each circle is assigned a number from 1 to 9,
$$a+b+c+d+e+f+g+h+i=45.$$
Then $a+b+c+45=3k$
$$a+b+c=3k-45.$$

Now, $a+b+c$ is at least 6 since the smallest numbers we can use are 1, 2 and 3.

Hence $3k-45$ is at least 6 and k must be at least 17.

The largest possible value for $a+b+c$ is 24, using 7, 8 and 9.

When $a+b+c=24$, we get $3k-45=24$ or $k=23$.

The possible sums along the sides are 17, 18, 19, 20, 21, 22, and 23.

45.   Problems of this type can be handled very nicely by using a diagram. We begin drawing an oval in which we indicate on what side of the river each of the participants is at any time. Hence (mg fw) means that the man and the goose are on the original shore while the fox and the wheat are on the opposite shore. The possible arrangements are:

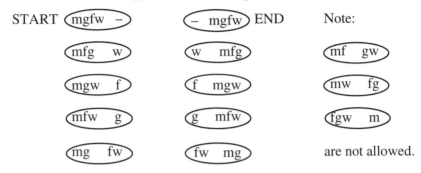

It is now possible to trace the moves necessary to get everything to the opposite shore, numbering the moves.

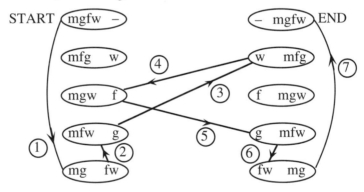

It requires seven crossings, which means that the man must row $7 \times 50 = 350$ metres.

Can you find another way to do this problem?

46.   I can see how to do this with seven parallel cuts, but that's not the smallest number of cuts.

The least number is not two because with two cuts I can create only four pieces.

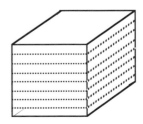

What about three cuts?
Aha! If I make three cuts, as shown, I
can cut the cube into eight identical
pieces!

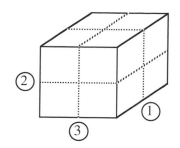

Can it be done in more than one way?

Well, if I make cut ① as before, and then
look at the front face, I can make cuts by
choosing A, B, C, and D all the same distance from their nearest vertex. Now if I
make cuts right through the block I won't
have cubes, but I will have identical pieces,
and I can do this in an infinite number of
ways, because A can be chosen anywhere
along the top edge!

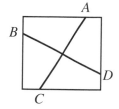

You might now be interested in seeing whether you can do this by
making four cuts. Five cuts. Six cuts.